# OCCULT ETHER PHYSICS

## TESLA'S "IDEAL FLYING MACHINE" AND THE CONSPIRACY TO CONCEAL IT

BY
WILLIAM LYNE

4TH REVISED AND EXPANDED EDITION

INCLUDES A LAST CHAPTER ON THE LYNE ATOMIC HYDROGEN FURNACE WITH UPDATES

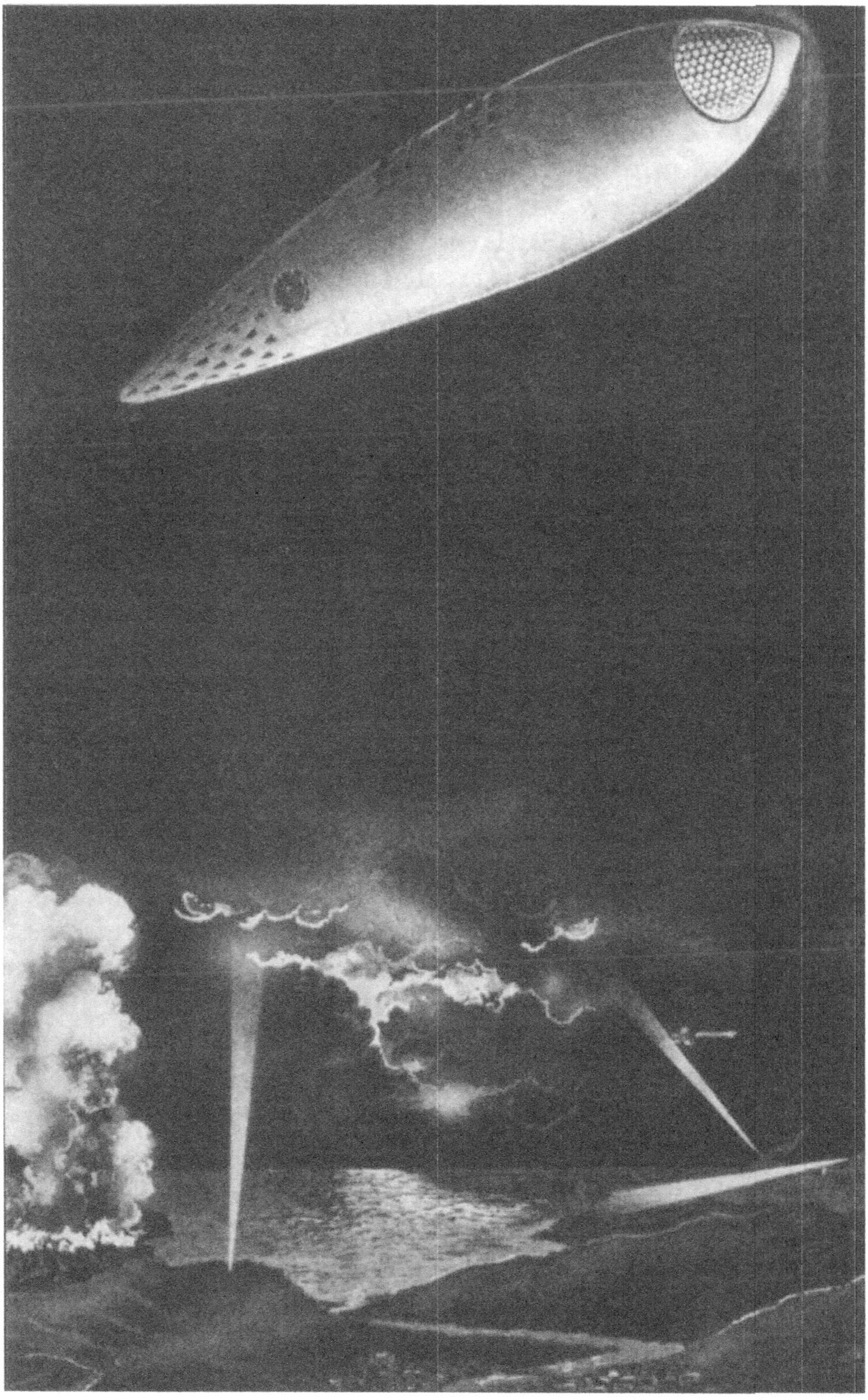

***The truth can flow from lies, but lies cannot flow from the truth.***

**Arthur Schopenhauer**

*Dedicated to my sons Angus Lyne and Matthew Olay*

Published by CREATOPIA™
William R. Lyne, 31-B Cerro Circle, Lamy, New Mexico 87540
U.S.A.
Tel./Fax (505) 466-3022

10987654321 Library of Congress
Cataloging in Publication Data

Lyne, William R.

Occult Ether Physics

I. Title

P.S. 93-90554
ISBN: 978-0-9854527-0-4

Non-fiction: A re-discovery of what the author calls "occult ether physics", the scientific/technical basis for a space-propulsion system invented in the late 19th century by Nikola Tesla, now used in the secret, exclusively man-made flying machines known as "flying saucers" or "UFOs". The government conceals this invention behind false scientific theories, "space alien" hoaxes, and false propaganda, disseminated through a controlled mass media. A "UFOlogy", "new-age", "paranormalist", and "pseudo-debunker" network is covertly maintained under "national security laws", to blur the facts of reality and disjoint inquiry, thus protecting the archaic technology and resources of coercive-monopolist, corporate-state interests from the competition of a truly free market.

Chapter VI is the Lyne Atomic Hydrogen Furnace, a 1964 free-energy invention of the author.

## TABLE OF CONTENTS

# PREFACE

**The two illustrations which precede the title page are ones commissioned by Tesla in preparation for his offer of inventions to the U.S. government in 1914, which was rejected by the government, so Tesla then made his offer to Germany to which he sold them. After all, a man is entitled to a reward for his work.**

**This revision was made necessary in order to get closer to the basic details in regard to Tesla's space propulsion technology. To this end, subtle changes were required. The previous 3rd revised edition has already experienced suppression which could only have come from the government through its catspaws. This revision therefore is to push the envelope even further than before.**

**There is much consternation today about the words "cult" and "occult", and what they mean. Many are worried that their own organizations, groups, or religions may fall into this category. Since the word "occult" merely refers to "secret" or "hidden", it is used by doctors to describe unknown bacteria as being "occult". As for a religion, philosophy, organization or other group activity, it is a "cult" if it utilizes "secret teachings", made known only to a select few initiates or "priests". Therefore, "national security secrets" are occult and hence the name for this book.**

**Is it possible that an entire area of physics has been declared virtually "occult" and "off-limits" to the public or to private individuals, by a powerful deep-black operations group, so that all science education or data generally available to the public has been "screened" to eliminate references to it? Is it possible that the actual science involved in "electro-propulsion" has been replaced with "bogus science", which has been promoted and promulgated to the extent that no one except the 'esoteric elite' knows the truth? And if so, has the "occult" science become the exclusive province of a very few specially selected corporate/government "scientific priests"?**

**I began studying this controversy—including the science involved —when I was but a child of eight years, in 1946, after interviewing oil field pumpers who had seen silvery "flying disks" in the remote oil fields in the West Texas and southern New Mexico desert. At the age of 13, I spent a couple of days visiting with family friends and**

their neighbor, astrophysicist Dr. Lincoln La Paz, in Albuquerque, new Mexico. La Paz was then employed by the U.S. government to help cover up Nazi flying saucer technology brought to New Mexico after WW II. My early convictions were later corroborated when I acquired a piece of hardware from a 1943 German flying saucer at Sandia Base where La Paz worked.

Since the truth about Tesla's space propulsion system is based on "ether" (A.K.A. "aether") physics, which has been collected, hidden away, or declared "national security secrets", or intentionally obliterated by false propaganda which is so intense that the general public remains unaware of the facts which are right under their noses, I have dubbed the ether physics ultimately connected to Tesla's system "Occult Ether Physics". This book is dedicated to the rediscovery of, examination of, and the integration of these secrets into the public knowledge.

The high priests of this occult knowledge are the elite owners of the coercive monopolist corporate banking, industrial, mineral and related interests, who in turn allow the initiation of certain scientists, military and government officials into their secret teachings only insofar as it is necessary, to execute and maintain a conspiracy of giant proportions, so that the technology can be classified as a "state secret" which can be controlled and enforced by means of unconstitutional "national security laws", military and civilian legal authority, and police. These measures are intended to protect the archaic technology and resources controlled by the elite, from the competition of a truly free market society. This technology is possessed by all the world's major governments, so we know that they aren't hiding it from each other, they are hiding it from us, the people.

Because of the last 76 or more years of progressively intensifying false propaganda and manipulation of knowledge and science, a majority of scientists ridicule the concepts of the existence of an ether and of flying saucers, as if such thinking belonged in the category of a "flat earth" theory and the "little green men" from outer space babble, but the reverse is true. The proponents of Relativism and the Big Bang want you to believe that the universe is finite, so that, if you go "too far" you will "fall off the universe", rather than off a "flat earth". What's the difference,? Flat earth...finite universe...they're both dumb theories.

And what do the "little green men" lies—originally told in 1947 by the government at Roswell—have to do with exclusively man-made electric flying machines, and technology based on sound and well-established electrical science and physics, but which, like the existence of the ether, has been tucked away in secret vaults, to keep you from using it, while you are being fed false and useless "theories"? These range from "curved space", "time travel", Big Bang", and "alien abduction" stories, to parallel universe", "inter-dimensional travel", "worm-holes in space", and any other outrageously incredible stories, concocted by government spooks, national security personnel, and secret agents in the "UFOlogy" and "pseudo-de-bunker" communities ("pseudo-de-bunkers" in my vernacular are official government spokesmen or scientific know-it-alls whose job it is to completely deny the existence of flying saucers, or to ridicule their other covert government counterparts, the UFOlogists, new-agers, or paranormalists).

Before you decide there is no ether, or no flying saucers, or before you believe the often-told "prime time" lies that aliens are "here", or that time travel is possible, you should ask yourself whether or not your thinking on these subjects has been conditioned by Big Brother's Biggest Lies. If you believe in relativism, I should point out that "time travel" is an integral part of that, which is a very incredible impossibility which makes the so-called "science" you rely on a worthless piece of pseudo-intellectual crap promoted by someone who wanted to sabotage your mind on behalf of the self-appointed high priests of "Occult Ether Physics".

Is your thinking a barometer of thought-control mediocrity? Are you so out of control of your own metaphysical concepts and epistemology, that you have to get your thinking from the "canned goods" section of NASA (NAZIA), courtesy of the CIA's Office of Scientific Intelligence ("OSI")? Is your education a frozen "TV dinner", prepared by a Secret Government, so you'll never discover the real food for thought?

For those of you who may be unfamiliar with my more complete exposition of the whys, wherefores, whos and other facts concerning the huge volume of false propaganda designed and disseminated on behalf of the coercive monopolist corporate state, to conceal flying

**saucer propulsion and other advanced and "free energy" technology, you should read my other books--- *Pentagon Aliens* (copyright 1993, 1995, 1999 (3rd Edition), ISBN 0-9637467-7-4, and *Occult Science Dictatorship* (copyright 2001, ISBN 0-9637467-8-2.**

**The flying saucer is a product of Nikola Tesla's life work, his most fundamentally important invention, for which he said all his other inventions were in pursuit of, though you have probably never heard this fact from anyone else. How do I know? I have documented it in many years of research, tracing Tesla's developments and his own words uttered and published during his lifetime, in scattered, sparse yet sufficient sources, as he placed a crowning achievement on top the work of the world's greatest scientific minds of the nineteenth century, who were Tesla's inspirations. And this invention has been concealed above all others by governments throughout the world. But they could not conceal everything, as there were already bits and pieces of information which were already widely published before they got their concealment act together.**

**It was supposedly proved, on May 29, 1919, "...that we live in a curved four-dimensional space-time". Albert Einstein said a massive body spinning in space "...would drag space and time around with it...", the origin of so-called "frame-dragging" and "space curvature". "Gravity is no longer a mysterious force acting at a distance but the result of an object trying to travel in a straight line through space curved by the presence of material bodies". (Gary Taubes, *Relativists in Orbit*, Discover, March, 1997).**

**Even such theories...relativistic bromides interlaced with such phrases as "the "fabric of space-time", "space curvature", and "frame-dragging", although off-the-wall ludicrous, suggest the existence of an ether, albeit an erroneous concept of the ether.**

**Since 1919, Nikola Tesla's propulsive discoveries have been concealed by government suppression, theft, and ridicule, while theoretical physics has become a pseudo-scientific mathematical shell-game which bears little new fruit, and less resemblance to reality. And this highly dubious "pop" physics is primarily what is spread by the media, while practicing physicists don't have much time for it. It is time for the truth as far as I am able to determine it.**

## INTRODUCTION: NIKOLA TESLA'S ORIGINAL 1875 PLAN

**Tesla's original vision, in his 1870s student days, of his "ideal flying machine", the realization of which is why he said he entered the field of mechanical engineering in 1875 in the first place; this was the problem which he had assigned to himself as his main lifelong work. (Frank Parker Stockbridge, The Tesla Turbine, The World's Work, March, 1912, pp. 543-48). The ship actually works on mechanical principles, but is actuated by electrostatic impulses. More on this later.**

**Originally anticipating that the electrical power needs for an electrostatically-actuated mechanical craft would be too great for an on-board power plant (since his ship was to be devoid of sustaining planes), he initially embarked on developing his "wireless transmission of power" system, so as to transmit the power needed from ground-based power generators and stations. The "power" being transmitted was actually mechanical in nature. Some of these earlier Tesla ideas turned out to be unnecessary for his flying machine, so he turned them for appropriate commercial purposes. Since his concept of the ether was of a relatively low-density one, and his earlier conception of the necessary force was not nearly so much stronger than the gravitational one as it later turned out to be, his later discovery that his anticipated electrical power needs were so much greater than that actually required, led to changes in his craft, and produced plenty of surplus "spin-off" technologies.**

**As stated in the Preface, in 1914, at a time when he was in great need of money, Tesla unsuccessfully attempted to interest the U.S. government in submarine and propulsion technology. In an article prepared by Tesla but never published, Tesla said "...we should push the development of flying machines and wireless transmission of energy without an instant's delay and with all the power and resources of the nation". The immediate association here is between "flying machines" (not "aeroplanes") and "wireless transmission of energy". Tesla had piles of patents which produced him no money, because the Morgans and Rockefellers had declared them off limits. No one could borrow money from the banks to put one of these systems into practice, and anyone who was foolish enough to do so**

**on their own money could expect the wrath of the gods to come down on them like lightening bolts from Zeus.**

**In 1914 Tesla sold his technology to the German Minister of High Marine, Admiral Alfred von Tirpitz. Tesla had also failed to interest anyone in the development of his wireless energy transmission system. He made the imminent discovery that his on-board power needs would not be so great, saying "...weight is no object". In 1913 Tesla had obtained a patent on a remarkable turbine he had invented about 1909, which solved his on-board power needs for his flying machine. This was probably in view of encouraging developments and awareness which showed that the electromagnetic interaction is $10^{40}$ times stronger than the gravitational interaction (George Gamow, Gravity, Anchor Books, N.Y. [1962], p. 138) According to Hendrick Lorenz, the electrical attractive force is 2x $10^{39}$ times stronger than the gravitational attractive force.**

**In an interview for a 1912 book, A Boy's Book of New Inventions - The Tesla Turbine, Tesla further revealed some details of his system when he stated the following:** ***"As I told you before when we were talking of the wireless transmission of power, the mechanism will be a development of the principle on which my turbine is constructed. It will be so tremendously powerful that it will make a veritable rope of air above the great machine to hold it at any altitude the navigators may choose, and also a rope of air in front or in the rear to send it forward or backward at almost any speed desired. When that day comes, airship travel will be as safe and prosaic as travel by railroad train today, and not much very different, except that there will be no dirt, and it will be much faster. One will be able to dine in New York, retire in an aero Pullman berth in a closed and perfectly furnished car, and arise to breakfast in London."***

**It was also implied that in producing the "ropes" the Tesla coil was used to produce rapidly-varying electrostatic potentials to create longitudinal sound waves in the ether for hundreds of miles through the air. The "equal and opposite reaction" to these waves was to bring back to the ship what are called "tubes of force" which carry mechanical momentum and, what Tesla's description appears to say, to lock onto a huge mass of ether molecules to pull the ship forward. As I witnessed on the ship I saw in 1953, an electrostatic envelope surrounded the ship as it moved without air friction, shock wave or sonic boom at around 36,000 mph. The description in the above**

quote by Tesla also sounds like a "tractor beam" technology. The principle of Tesla's turbine was to use the adhesiveness and viscosity of rapidly moving gases to propel his turbine blades around. Applying this same principle to his aircraft would use the adhesiveness and viscosity of the ether applied through a steady brush discharge, to impart momentum to pull his ship through space.

There was a change in Tesla's strategy which seems to have occurred between this 1912 and 1915 tests of an electro-mechanically powered missile using wireless energy projected to it by a power beam from the ground. These new developments were probably spurred on by necessity—the "mother of invention"—since without a wireless energy system, the prior conception would be rather useless. The power monopolists could fight that off forever, so Tesla was left to his own devices, unable to count on anyone else. Typical for a true genius, when he was thwarted in one direction he found another way.

The earlier conceptions must have been still in Tesla's mind when he developed the turbine commercially under the Tesla Propulsion Company slogan, "20 Horsepower Per Pound", with the intention of also using the turbine to drive one of his specially designed high frequency alternators to power his ideal flying machine. After having developed such a light and powerful turbine, which could turn upwards of 30,000 rpm., with a ten pound unit producing 200 break horsepower on gasoline, he also designed and received a patent in 1928 for a new kind of "VTOL" (Vertical Takeoff and Landing) -type propellor-driven airplane, to be powered by two turbines mounted in tandem, which would counteract the torque of the prop, and avoid the need for the stabilizing rotors that helicopters have on their back fuselages. This aircraft, if built, would have had a superior weight-to-horsepower ratio—perhaps three-to-one—and would have been fast in level flight, taking off vertically.

On seeing the inefficient heavier aircraft engines then in use, and the requirements of more cumbersome aircraft with greater wing area but lower wing-area-to-weight ratios, Tesla apparently wished to commercialize on the turbine by designing and patenting the new airplane, which specifically used his turbine, to raise needed capital to complete research and development on his more expensive flying machine project. In his experiences with prior patents, he must have learned that a patent would be worthless unless he could get the

**needed capital to carry the project out before the patent period expired, and that the disclosure in a patent would allow others to do so only 17 years after a patent might be granted to him, during which time his invention could languish for lack of funding on his part, because the monolithic financial interests were set against him. And yet there was the patented airplane and no one built it.**

**While in the initial stages of development, Tesla seemed quite unaware of the true power of "the force", until his early field tests proved that it was "tremendous". In the years 1893 and 1894, Tesla had "completely worked out" his Dynamic Theory of Gravity", that is, at least in "theory". Inherent to that theory, was his propulsion system, yet in the formative and theoretical stages. To publish his theory however, would have been tantamount to disclosing the basic technology as well, which would have allowed his competitors with more money and resources to complete and take credit for the most important work of his life, and the greatest invention of mankind. Tesla lacked the necessary money, and could not raise it. He could not protect himself by patent, until he completed and tested a working model of large enough scale to impress the aircraft industry and the government. Fatefully, it was during this time of greatest financial need, when he still had the physical stamina to carry the project out, that he was run ragged by bill collections and foreclosures, at the behest of J.P. Morgan and his cohorts, all with the intent of crushing Tesla because he kept coming up with new electric power systems which made his prior systems obsolete. Since the newer systems would have ruined Morgan and Rockefeller, and since any invention which Tesla marketed for great profit would have allowed him the resources to build more energy systems, he had to be crushed. Tesla never seemed to fully realize that Morgan's pseudo-philanthropy became nothing more than a scheme to compromise Tesla's contract with himself, and a way of falsely inducing his reliance on others rather than on himself. This treatment set Tesla back emotionally as well as financially, after he had invested all his cash in his project, and reduced him at one point to a heap of groveling nerves, literally begging Morgan to make good on his promises, to no avail, but like all truly tragic figures, there was a peripides—a reassertion of the tragic hero's excellent and untarnished Arturian powers—before he passed heroically into the**

**oblivion of death, as his discoveries, like the sword Excalibur, sank beneath the waves of the lake of 'national security' lies, to the inurement of corporate greed, coercion and extortion.**

**The only power that Tesla seemed to have left, was that of withholding his discoveries from the world, to protect his secrets from the pilfering paws of the looters, until and unless he could receive the compensation and credit which he deserved, but that didn't matter, because they got it all before his corpse was cool, and never paid him so much as a dime for his trouble.**

**After becoming aware of the true power of "the force"—one hundred billion thousand trillion thousand trillion times stronger than gravity—Tesla began to pursue the idea of an "on-board" power system for a manned craft to be used for personal and commercial transportation, with his "wireless transmission of power" to be relegated to a robotic, remote controlled, electric powered transmission of mechanical power to a "missile" from the ground, for use in warfare. After a long period of silence following his field testing of a working model of the electro-mechanically propelled missile in 1915, Tesla began referring to the propelling force as "tremendous" in his later announcements which began in the '30s. "Tremendous" does not mean something which reacts feebly, to barely get off the ground, but to something which rises with great and astounding force, such as the way I saw a flying saucer perform in 1953...only ten years after Tesla's death. If "tremendous" meant something like an airplane or conventional rocket—as the CIA/NASA mis-informers have idiotically attempted to misconstrue —then what would be the word to describe a force which propels a flying saucer at 36,000 mph in three seconds?**

**The words "means of propulsion" used by Tesla in his 1940 interview with William L. Laurence of the New York Times are of the utmost importance for those interested in understanding both the misinformation conspiracy as well as the scanty documentation of Tesla's concealed discoveries. In their context, these words denoted something which was different from the autos, planes, trains and ships then in existence, else why mention them and why would it be "new"? These words obviously meant an on-board electro-mechanical means to be used on his "ideal" flying machine, either manned or unmanned, controlled "mechanically" by a pilot on**

board, or "remotely by wireless energy" by a controller on the ground. It did not refer to a "projectile" such as a "pellet" or "bullet", which had no "propulsion system", and was merely "repelled" by a gun of some sort, and it did not refer to any kind of aerodynamic "airplane"—hence Tesla used the term "flying machine"—since his machine was to have no wings, ailerons, propellers or "outer appurtenances of any kind". "Propulsion" here means an on-board system for perpetuation of motion by electro-mechanical means supplied with electrical power by either on-board turbine-driven generator or by mechanical energy and control signals transmitted by power beam from the ground. The idea of interplanetary travel appealed to Tesla, with the idea that, so long as the energy for propulsion could be transmitted from the earth, a space ship would require no fuel tanks. But the term "veritable ropes of air" raises a question as to whether these ships will work in outer space.

In a letter to his friend and financial supporter B.A. Behrend in the 1930's, Tesla referred cryptically to what may have been his propulsion discovery: "What I shall accomplish by that other invention I came specially to see you about, I do not dare to tell you. This is stated in all seriousness". Tesla may have been inferring a plan to discover "other worlds" in outer space, if some version of his flying machines would work there. We really can't know for certain how all of Tesla's secret inventions worked but the government knows. There seemed to be some differences between Tesla's earlier statements and later ones which indicate that he had several different related inventions and that he may not have understood himself exactly how some of them worked.

Tesla said that all bodies contain "electrical content" and that they behave as resonant bodies which interact with rapidly varying electrostatic forces and ether to determine their gravitational interactions and movements in space. These theories were tested and confirmed to some degree in his 1899 Colorado Springs experiments. The U.S. government never gave Tesla the time of day, and his great propulsion discovery slipped into the eager hands of the Nazi elite, when Wernher von Braun engaged and began development in the "p2 Projekt" at Los Alamos, New Mexico in 1936. Tesla apparently participated—in fact may have been the project leader—in these early tests in New Mexico from 1936 to 1938. Von Braun took the data back to Germany in the Winter of 1937, where research and development was conducted at Peenemünde in the Baltic, with underground manufacturing facilities elsewhere in Germany, until these secrets and others were traded to American corporations and the U.S. government in exchange under Operation Paperclip, just after the armistice, for amnesty for many war criminals, along with an agreement for several thousand people to work for the U.S. government and American corporations. Anyone having important technical information was protected. The Nuremberg Tribunal was a show trial which did not prosecute anyone having advanced or "strategic" technical knowledge or abilities even though they may have been guilty of crimes against humanity.

I have always associated the ship which I saw up close in 1953 with Tesla technology, because of the copious electrostatic discharges (which I call "Tesla discharges") because it was obviously an electric flying machine, and because the ship's precession indicated gyroscopic stabilizers as Tesla emphasized. And because the ship moved with a force consistent with the fact that the electromagnetic interaction is 1040 times greater than the gravitational interaction.

In 1912-14 Tesla mentioned "veritable ropes of air" which "pulled" his ship through space (there being no "thrusting" from behind) and said that his ship utilized the same phenomena—the adhesiveness and viscosity of fast moving gas—as that which propelled his smooth turbine blades around, except in this case the "gas" he referred to was the ether. He stated that his ship or a missile based on his system could be "...dropped on any point on the globe" electrically. Is it possible that Tesla stumbled on something much more powerful when conducting tests, attempting to use air? Is it possible that J. J. Thomson's description of the specific action of the technology involving the ether is more accurate (that "...the aether is itself a storehouse of mechanical momentum")?

There are those who have interpreted or will interpret Tesla's statements as meaning that his flying machine technology was simply a matter of using his turbine and Tesla pump to suck air in and shoot it out the rear of the machine to produce the thrust to "repel" it like a "jet" engine, yet we know this is wrong because of Tesla's use of the word "pull", and there is no way to create a "rope-like" column of air with only his turbine and pump powerful enough to "pull" his ship. And there is certainly no technology based on such a primitive idea which could accelerate a ship to 36,000 mph in three seconds as I observed, with no sonic boom or shock wave. And with no intake ports or jet nozzles on the rear. But there is a technology based on Tesla principles which could and did do so electrostatically, and that technology appears to have been more cogently described by J. J. Thomson. And it is also possible that what J. J. Thomson described was partially technology which Tesla was attempting to keep secret.

Thomas Edison joined the Theosophist organization of Madame Blavatsky in 1872. Their main activity seemed to be industrial espionage, concealed behind a veil of an interest in mysticism. I think Edison used the organization to steal technology from other inventors. Tesla was invited to join in 1895 and he refused. His laboratory was burned to the ground soon after his refusal. J. J. Thomson was also associated with the Theosophists. Was some of the technology he described obtained from Tesla through Theosophist espionage?

## The REAL Cover-up

Ever since Tesla's saucer discovery came under the control of the U.S. government in 1945, there has been in effect a false propaganda program—originally developed by the RSHA VI—Reichsicherheitampt VI, or "National Security Agency Division 6"—the division of the Gestapo which was invested with the highest secrets of the German Reich (managed by General Reinhard Gehlen) which included and controlled in Nazi Germany the equivalent to our own UFOlogy, new age—and what I call the "pseudo-skeptic misinformation dissemination network." The U.S. government simply employed the whole Nazi saucer counterintelligence apparatus, and it was some of the original Nazi personnel employed at Holloman, A.F.B., at Alamogordo, New Mexico, who designed and implemented the 1947 "Roswell Hoax". Though shadowed with ostensible military and "national security" vestiges, this cover-up has been protracted on behalf of a coercive monopolist corporate state—even in Germany—ever since the acquisition of the secrets by the

**Nazis, for if the technology had gotten even into the hands of private Germans at that time, the Nazi state's days would have been ever so much more shortly numbered.**

**Can you imagine the effect of this kind of transportation technology in private hands? It would allow the common man to go anywhere in the world, to visit all the places which he cannot presently afford the airline fare to. Yet a small, privately-owned flying saucer would cost less than an automobile to build, would use practically no fuel, would travel thousands of miles per hour, would last a long time, and would require practically no maintenance.**

**Can you imagine the effect it would have on totalitarian communist, fascist or corporate state governments and 'privileged' economic interests? Can you see the bottom drop out of the over-priced real estate market, when it suddenly became possible to live anywhere, even in places where there are no roads? Can you see the problems along the borders of countries, if saucers could fly across and back without checking through "customs"? (The conservatives and totalitarians are horrified by this possibility, as if our borders do not already leak like sieves.) The main problems, as you can see, are with the paranoid and criminal fears of power crazy authorities and coercive monopolist industrial interests, who would suddenly lose a lot of the power necessary for them to keep their "jobs" protecting and maintaining the interests of themselves and the coercive elite of whatever kind.**

**These considerations are the ones which indicate the perpetrators of any "cover-up". There is much talk on TV programs, radio talk shows, newspapers, books, magazines and videos about a "...cover-up of our government doing business with aliens for fifty years", etc. But if you can see it on prime-time TV every night of the week, the only "cover-up" which can be presumed, is one which is opposite to the one they pretend to be telling you about. Such programs are the cover-up. The volume of these lies is so great that it is almost incredible, considering the fact that they are evidence of our own government covertly acting behind the scenes to lie to us. Whenever government is involved in broadcasting such trashy BIG LIES to every household in America, it is good cause to ask, "What else are they lying to us about?"**

**This book, along with my prior ones, are dedicated to the**

**objective of bringing this technological secrecy and socio-economic dictatorship to an end. We are being robbed of our future. This 2010 Edition was called for because between the Second Edition of 1998 and the present, more information has come to me, but the original nature of the cover-up has never changed since 1947.**

## CHAPTER I: THE OCCULT ETHER THEORY AND ELECTRO-MECHANICAL PROPULSION

**The theory which I espouse in this book is based on what can be corroborated throughout the words and developments of Nikola Tesla and their logically construed implications, in combination with other special information linked to that, as presented here and in my prior book, Space Aliens From the Pentagon (now in its Third Edition as PENTAGON ALIENS). This theory is close to Tesla's Dynamic Theory of Gravity, and I like to think that it even suggests some corrections that Tesla himself made or would have made if we only knew what is concealed in government vaults.**

**Like most theories, this theory is incomplete, may contain errors, and will require further development, tests, corrections, and revisions, either by myself or by others in the future. I have stated here my basic thinking on the subject up to this time, which has progressed somewhat since my prior books.**

**Since this book is for the greatly uninformed public, as well as for the researcher, I will make an effort to couch it in laymen's terms —for I am myself a layman—and will provide some illustrations where helpful, to give a reasonable grasp of my concepts based on Tesla's concepts which I have developed as a consequence of what I have also witnessed. I do this rather than engage in a "quantum mechanical" mathematical orgy, which is usually combined with pages of dull equations and numbers which proceed logically from the contradictory and unproven illogical premises which are so prized by that ilk. It is as if to say, "Hey! Look at me Ma, I can do lots of math!"**

**"Ether Physics" by its very nature and name is subatomic physics, which to me means "sub-protonic" and "sub-electronic". I am not so concerned with sub-atomic nuclear particles per se, but**

**rather with what I believe are the finite building blocks of the proton and electron, which exist within the ether in simpler form. I will leave all the quarks, muons, mesons, leptons, etc. to the "Relativistic Quantum Mechanics" (hence "R.Q.M.s") and their "elegant equations".**

## Characteristics of Basic Ether Particles

**The basic ether particle, on order to respond to the negative electrostatic D.C. Potentials described by Tesla, must have a positive nucleus—a "protette"—and a negative sub-electron—an "electrette", and is surrounded by an insulative fluid as Tesla said. As you may have noticed, this scheme is a scaled-down version of the basic hydrogen atom, with its proton and electron. Like most atoms, it is normally neutral and in equilibrium, but is much smaller, being ultra-fine. Due to its tiny size and neutrality, it can pass easily through solid bodies which it permeates as it moves, yet the ether as a whole behaves like a solid in respect to rapidly-varying electrostatic potentials of specific frequencies, from the infra-red through the visible light frequencies, which disturb its equilibrium, yet, though we can feel it, it appears to be transparent and invisible to the naked eye.**

**Like hydrogen gas, there is some elasticity, so it is "elastic" as Tesla said, but is incompressible. Due to the tiny size, so called "empty space" is virtually packed with this very fine matter, which oscillates at such high frequencies—well beyond that of x-rays—yet the tiny size and normal neutrality allows it to penetrate "solid mass"—which is mostly "space" (which is also saturated with ether which must be moved through by a mass). And as J. J. Thomson said, "The ether is the storehouse of mechanical momentum".**

**Since interstellar space is equivalent to a vacuum containing little gaseous matter of the conventional type—such as a Lenard tube in which charges easily move—ether-filled interstellar space is permeated by charges moving freely along frozen magnetic lines of force as Hannes Alfvén said.**

**Ether-packed space is also omni-directionally interpenetrated by ultra-fine radiation, which Tesla called "Primary Solar Rays", which Tesla said were positive particles propelled by enormous**

repulsive forces in stars. I think these Primary Solar Rays may constantly "twang" the electron clouds of atomic matter, creating or powering "electron orbits", which may be in fact standing waves in the undulating layers of particulate electronic clouds surrounding atoms and molecules. Since the matter with which these electron clouds are associated is in constant motion, the clouds are currents, since a "charge carried around is a current". The driving force behind the currents appears to be the motion and the Primary Solar Rays, also by a conversion of weak forces into strong ones. Any lost momentum is 'made up' by the Primary Solar Rays and the conversion of weak to strong forces, the ultimate source of extra force ("free energy") needed to make up for the losses in the universe. The conversion of weak forces into stronger forces will do more work in the same period of time. The Primary Solar Rays are an example of this process. With the UFO, the gravitational, kinetic force of a turbine propelled by chemical reaction is converted into electrostatic force which is much stronger than gravity. This is free energy.

The ether must have an almost balanced charge-to-mass ratio, and respond to both positive and negative electrostatic potentials. Tesla said the positive or negative aspects depended on the conditions. Electrostatic generators provide a good example of the electrical behavior of the ether. At sufficiently high voltages the positive and negative dynamic sub-charges are separated by magnetic fields and condensed as electrons and "holes". There is some unknown underlying reason why these charges are of specific magnitude, which is yet to be explained, but the quantum theory—originated by ether physicist Max Planck—practically admits that there are subdivisions of basic electrical charges, and that electrons are not "indivisible". This is one basic bone of contention, since if electrons are divisible, Relativism must fall. I think the "basic charge" which has been perceived as that of the electron, appears to be time-related, since electrical current moves virtually at the same speed as light, therefore my opinion is that the electron charge is probably the quantity of charge accumulated from a discrete number of negative sub-charges carried by ether over some discrete unit of time due to the positive mass units of ether composing a proton. This may in turn reflect the distance traveled through space of a proton during that time, with the charges circulating as a current between the ether and grosser matter.

High voltage force—either in electrostatic impulses or in radiation—is apparently necessary to force the ether to form momentum-carrying tubes of force, as an 'equal and opposite reaction'. The principle applies to Tesla's propulsion system, which was a sort of electro-mechanical one. Strong, high voltage, negative charges are necessary to penetrate the insulating fluid surrounding the ether to react with the positive ether mass to cause the "inertia resistance" of the ether to be "asserted" as Tesla said, so as to access its mass—and that of the atmospheric gases locked to it—to be possibly pulled against by the ship. The screwing force of tubes of force which rotate around irrotational vacuous ether cores may be the "positive mechanical action" referred to by Tesla and the "pulling force" involved. Unless the tubes of force are dissolved in the conductors of the ship imparting momentum to it, as J. J. Thomson said (more on this later). The tubules could be the product of the rotation imparted by a magnetic field to electrical currents, which are consolidated to effect changes in momentum.

The rapidly-varying electrostatic forces emanating from the earth rigidify the ether within the earth's electric field, which may form a virtual stationary anchor which an electro-mechanical craft may possibly lock onto to propel itself through space. Although the ether 'anchors' are stationary relative to the earth and carried along by the earth's electric field, this carry-along ether has a velocity of many thousands of miles per hour relative to the ether outside the earth's electric field. Just as the so-called "gravity field" diminishes as the square of the distance, the relative movement of the ether decreases outside the earth's electric field.

If gross matter is formed of ether particles, what is the "mortar" which holds them together? Tesla said the ether was particles within an "insulating fluid". This fluid is apparently what 'wets' positive bodies and particles as a 'perfect fluid' which flows everywhere. Does it have viscosity and adhesiveness? If it does then forcing the incompressible medium past or through the ship could be the key, sort of like swimming under water.

While the ether is invisible to the eye—because its period of vibration is beyond the visible frequencies—it is not imperceptible.

**As we move, we can feel the ether accelerating, decelerating, or changing directions within our bodies. When running, we feel the resistance ("inertia") created within the atoms of our body, as a response to changes in momentum, as the tubes of force are forced in different directions or in accelerating or decelerating pathways through the atoms and molecules of our body. Once we reach a constant velocity, we can feel how the tubes of force 'bore' through our mass, delivering the momentum the ether carries, which tends to carry us along in level and straight lines (the paths of least resistance). Once in constant velocity we must supply only whatever momentum is shed through wind or mechanical friction or erratic movement, in order to maintain constant velocity, and we can feel all this. We can also see the effects of the ether in the behavior of everyday objects—the way they resist changes in movement, fall, accelerate, gain momentum, tumble, roll, collide, shatter, bounce, or come to rest. I analogize a solid mass to a flock of geese, maintaining its "formation" ("shape") as it flies trough the air (ether) as a single unit:**

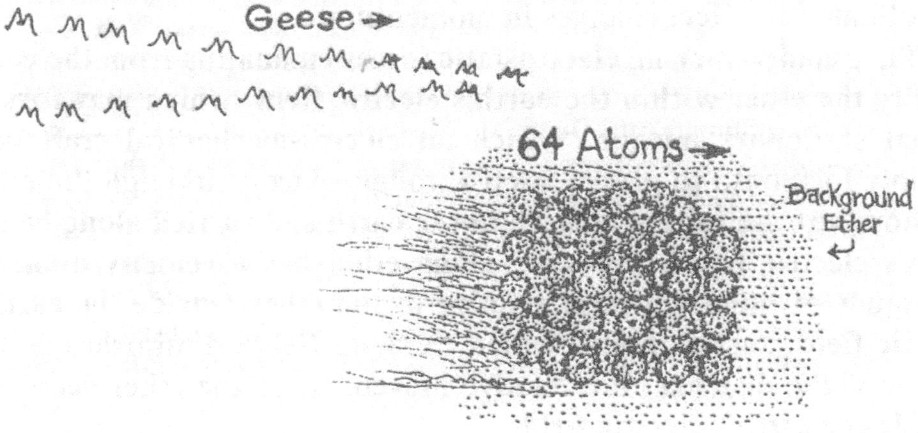


**Newton's laws fail to tell us "why" gravity exists, what forces create it, or why the force varies "directly proportional to the masses, and inversely proportional to the square of the distance between (the centers of the) bodies". If gravity is a force at a distance, why would a body in outer space be completely free of it? Why would the force of gravity between the earth, for example, and the moon, not affect a body floating in outer space between the two? If that were true,**

would the force selectively pass through the body without affecting it? This fact negates Newton's laws. Even Newton explicitly admitted that his laws failed to address causes, and only claimed to mathematically describe "apparent effects". But what may "appear" to be true may not be all or any of the truth.

"Electrical content" is inherent to all solid matter. Atoms contain nuclei composed of positive protons and neutral neutrons (neutered by the tightly bound electrons within them). Around these nuclei are negative electrons. All matter which contains electrostatic charges and is in motion, represents "electric charges carried around", hence currents, which create magnetic fields. Since magnetic fields impart rotation to electromagnetic fields, this creates rotatory electromagnetic tubules within the ether within the fields. These tubules rotate around irrotational vacuous cores in space and within solid bodies relative to their motions, and are aligned by the relative motion along the axis of a uniform or changing movement, and impart momentum when dissolved in a body. While at any given time the pitch and diameter of a micro-helix is fixed, it involves a logarithmic spiral or vortex either inward or outward each time the velocity changes, and the pitch resists, then adjusts to the new momentum or direction. But with Tesla's system this polarity of movement can be changed electromagnetically, so that all the internal reactions can be instantly reoriented, thus eliminating such things as centrifugal force and inertia, created by the tendency to remain in a particular state of motion.

According to Tesla, the earth and all electrostatically charged bodies have a maximum negative surface charge of approximately 20 esu/cm2 (20 electrostatic units per square centimeter), an atmospheric phenomenon due in part to the ionizing effects of cosmic radiation and lightening. The relatively non-conductive, insulating and dielectric properties of the atmosphere play a part. Beneath the earth's surface charge, the semi-conductive crust contains a vast ocean of surplus electrons. The earth's surface charge is more or less a constant, due to the earth's surface area, varying somewhat by altitude due to the changes in atmospheric pressure and moisture content of the air, as well as the incidence of cosmic radiation and its frequencies. This charge leaks upward constantly but is maintained by lightening.

The nuclei of all atoms—including hydrogen—possess net positive charge. This is because the basic positive charge of a proton exceeds slightly the magnitude of the basic negative charge of the electron. Hydrogen, with its single proton and single electron, has the highest charge-to-mass ratio and the lowest gravity of all elements. As one moves up Mendeleev's periodic table of elements, progressively increasing in complexity and mass number above hydrogen, elements possess progressively lower charge-to-mass ratios, since the comparative ratios of electrons match the number of only the protons in the nuclei. Yet since the neutrons are not negative, they add to the positive mass relative to the atom, but contribute nothing to effective electrical charge, and the ratio of neutrons and mass increases up the chart. In relation to the earth, this leaves an excess positive mass, the semi-conductive surface of which is capable of receiving and maintaining a negative charge surplus, mostly brought to earth by the thousands of lightening bolts per second, and retained due to the dielectric and insulating properties of the atmosphere.

The constant electrical discharges from the atmosphere by lightening are driven by the cosmic rays and motion of the earth, which converts the electrostatic charges to currents. This was confirmed by H.A. Rowland's experiment. Since atmospheric gases are insulators, the mostly alkaline, semi-conductive crust of the earth is largely an electron donor material, which maintains a large reservoir of negative charges insulated by the atmosphere. Cosmic rays striking the rarified light gases of the ionosphere, maintain the ionosphere's positive charge. The excess negative charges of the earth are meanwhile ejected constantly by the rapidly-varying electrostatic discharges (the "microwaves") which Tesla discovered. The effects of gravity also helps, since helium and hydrogen, both light, electropositive gases, gravitate to the ionosphere. Between the ionosphere, ca. 620 miles above, and the earth's surface, there is a gradient of ca. 150 volts/meter, totaling about 176 million volts, creating a considerable electric field, but the electric field extends far beyond the ionosphere. This electric field creates electrical displacement in the ether within it. The effects of the electric field on the ether is almost instantaneous, since near the speed of light, as the ether passes from "free space" into mass, where the gravitational force is exerted downward, toward the source of the electric field. The relative weakness of the so-called gravitational force however might be due to bodies on earth being pulled upward due to the downward motion of the tubes of force being brought down by the rapidly-varying electrostatic discharges from the earth.

There is no significant gravitational effect of earth's field above this electric field. Since earth's magnetic field encompasses the moon, so does the electric field. When the ether is placed within a strong electric field, it is polarized and the negative charges on it are attracted by the positive pole (ionosphere) and repelled by the negative pole (earth). The effect of these attractive/repulsive forces is to "displace" the ether, and to place the dielectric atmosphere and the ether within a strain. Tesla and other ether physicists said this strain was the cause of gravitational effects.

## The Cause of the Apparent "Inverse Square" Law of Gravity?

As earth spins, the surface velocity—and the atmospheric velocity with it—at any point varies according to its distance from the earth's axis. An object sitting at sea level moves slower than a body on top Mt. Everest. The voltage potential between the atmospheric gases at higher altitude and the ionosphere is lower, but the voltage potential between the atmosphere and the surface of Mt. Everest is higher, which, along with lower air pressure explains the increased coronal discharges on mountain tops.

The differences in these electrical potentials illustrates why there are differences in relative displacement of the ether within both earth's "gravity field" and within bodies moving within that field, with a degree of displacement which is directly proportional to difference in so-called "gravitational force" and the dielectric strain.

This comparison suggests that a body at a higher altitude and velocity has a lower gravity because the electric field has a lower displacing influence on the ether . This is pertinent to electro-propulsion because we must diminish, reverse, magnify and otherwise control this displacement in order to instantly synthesize and control inertia, momentum, and "gravity". This displacement influences the momentum because it affects the vibrating micro-helical tubules and the way they "bore" through etheric space.

"Gravity" appears to be a product of the resistance to a change in ether displacement. The reason why a moving sphere when charged experiences an increase in its "virtual mass"—that is, an increase in the work required to move it, but not an increase in its gravity—is because the "Faraday cage" effect of the increased charges tends to increase the resistance to the movement of the mass through the ether, apparently effected by the negative surface charges on the sphere.

## Tubes of Force

Due to the cosmic motion of earth, electrostatic, molecular charges become currents, which create magnetic fields within atoms, molecules, and sometimes ponderable bodies such as iron. As stated before, as earth moves through the ether, rapidly-varying electrostatic impulses from the earth rigidify the ether within earth's electric field, causing the ether within the field to be carried along.

The tubes—what Tesla called "moving molecules"—may function like drills, tending to bore straight paths at a constant pitch and linear velocity, around irrotational vacuous cores (consistent with Newton's laws for inertia and momentum). Any change in direction or velocity would apparently be met by a resistance ("inertia"). Since all mass is already in motion, it always has momentum, acquired from the ether. The inertia exhibited by a body at rest relative to earth is apparently a product of its momentum, so this inertia is really momentum. Inertia is therefore a resistance of a body to change in its state of momentum. The two words are used in different ways, to define "relative changes" in velocity and/or orientation, but no body is "at rest".

Since the ether near the spinning earth is subjected to the negative impulses (the "current") and the positive mass of the earth, the ether forms tubes of force which are attracted downward. With the theory of the moving tubes of force a mass acquires momentum from the tubes entering it. Under this reasoning, a body sitting at relative rest on the surface of the earth would be attracted upwards. So what attracts the body downward? Could it be that the momentum through space that the body already has is somehow being re-directed downward by the earth's electric field? Could the momentum from the tubes of force which carry the body through space along with the earth be partially deflected downward by vertical tubes of force? If gravity is the product of a body's reaction to the ether's resistance to a change in its momentum, transferred to a mass due to its constantly changing movement, orientation and velocity as it moves through universal ether-filled space, gravity could be explained. The explanation provided by Sir William Thomson and Nikola Tesla are given later in this book.

## More on Tesla's Primary Solar Rays

According to Tesla, in addition to the ether all space is omni-directionally interpenetrated by ubiquitous "Primary Solar Rays", particles repelled by positive forces, causing extremely high velocity and energy. This is electrostatic inductance, using repulsive rather than attractive forces. These rays are the product of particles which are projected by the enormous voltages of stars, including our sun. They may be similar to what contemporary physicists have called Zero Point Radiation ("ZPR"). One scientist has attributed so-called "blackbody radiation" to the ZPR (T.H. Boyer, Derivation of Blackbody Radiation Spectrum Without Quantum Assumptions, Phys. Rev. 182 [5], pp. 1374-83 [1969]). Boyer also found that a nonlinear dipole can absorb energy from particular modes of the ZPR spectrum (Random Electrodynamics: The theory of classical electrodynamics with classical electromagnetic zero-point radiation, Phys. Rev. D 11 [4], pp. 790-808 [1975]). Another scientist claimed to measure the energy density of ZPR at 1094 grams/cm3 , for individual fluctuations of random ZPR, on the scale of the Planck length (10-33 cm.), by means of a standard radiation detector, the frequency response of which was specially altered to extend into the super high frequency spectrum involved (J. A. Wheeler, Geometrodynamics, Academic Press, N.Y. [1962]). In my final chapter, I detail what I think is the extraction of Primary Solar Ray energy in the atomic hydrogen process.

These particular researches of 1962 to 1975 are ex post facto, coming long after Tesla's great discoveries. They merely confirm that, in order for such intense radiation to pass through space packed with an even finer ether, probably consisting of ultra-fine bundles of

**positive mass, surrounded by negative charges and an "insulating medium", as well as passing through other planets and the earth, it would require the enormous force that Tesla described which is generated in stars like our sun, propelling tiny particles which do not react with the ether in space, and only react with certain materials which we call "radioactive materials". This reminds me of the multi-frequency response of interlinked tetrahedronal silicon dioxide crystals which make up solid quartz crystals, which can be cut to respond to a wide variety of frequencies.**

**If the Primary Solar Rays pass through space and mass—which we know is mostly "space"—and they react with so-called "radioactive materials" as Tesla said, what is it about the structures of radioactive materials which causes them to capture the energy of the Primary Solar Rays and to emit radiation? And if so, wouldn't the development of technology to convert the force of these particles to usable energy be an excellent way to supply all our energy needs?**

**I have previously mentioned a technology which can use radioactive emissions to ionize air and to separate high-energy alpha and beta particles with a magnetic field to produce a huge output of electrical energy.**[1]

**In reiteration, Tesla's propulsion utilizes externally projected electrostatic impulses of high voltage and frequency to "rigidify" and cause the ether to assert its "inertia resistance", combined with (polarized) rotatory electromagnetic tubes of force, which instantaneously impart momentum to propel a spacecraft.**

## An Old Wooden Box and a German Inertial Guidance System

**In 1979, at a Santa Fe flea market, I examined an old wooden**

---

1[1.] *PHYSICAL CHEMISTRY,* E. A. Moelwyn-Hughes, Pergamon Press, Oxford, 1965. Page 224: "Rutherford and Geiger found the number of α-particles ejected per second from 1 g of radium was $3.4 \times 10^{10}$ ($k=1.26 \times 10^{41}$ sec.[4], later determined by Madam Curie at $1{\cdot}38 \times 10^{-11}$)". Each particle carries two units of positive electricity. This will produce 34 zillion α-particles per second, at 4.5 Mev. The number of charges per second, yields a hugh amperage which, when stepped down to normal 120 VAC, is staggering. Page 230: "1 atom of radium expels one α-particle producing 1.5 billion ions, at 4.5 Mev". Page 231: "One α-particle has more than sufficient kinetic energy to bring about all the ionizations which stand to its credit", and is "...quantitatively sound".

**box belonging to an elderly Pojoaque man who had been employed as a workman in 1937 at a secret project called "p2" at nearby Los Alamos, which was run by Nikola Tesla and Wernher von Braun. The man had built the box to carry tools used in his job there. He had carved in "New Mexico folk style" a "map" on the lid of the box showing the highways, the location and name of the project, the symbol for the project (a triangle with a dot in its center) which was apparently a symbol for the Prussian/Masonic "Illuminati" of which von Braun was a member, and the date of the project. This information was corroborated by my friend Peter van Dresser who confirmed that von Braun was involved in a German research project at Los Alamos at that time. Pojoaque, where the old man lived, is prominently shown as a large "P" with a zia sign around it. The zia sign is an ancient symbol used by the Zia American Indian tribe to denote the sun, which they worshiped:**

**When von Braun arrived from Germany he was accompanied by 16 top Nazi scientists. Dr. Robert Goddard, who shared American Rocket Society technology with von Braun, purportedly left the**

project and returned to his workshop at Mescalero when he realized that von Braun's interest was in Nazi military applications. Von Braun was able to acquire the new rocket engine design from the Rocket Society's publication, *Astronautics*, which he acquired from Dr. Robert Goddard.

(NOTE: There has been no publically-available information whatsoever on this project and those at Los Alamos who later conducted the Manhattan Project apparently had no idea as to what the old barn used by Tesla and von Braun in the "p2 Projekt" was for, or that it had an historical significance, until my first book was published in 1993. The barn had been rented from Spanish Land Grant heirs by the p2 Project about 1934. Around 1979, LASL, not knowing what it was used for, had cleaned out the contents of the old barn, removed all the copper wire used on the coil forms, and sold the contents to salvage dealers. It was obvious that the old wooden forms, which were mostly rough sawn planks connected together with dowels, were Tesla coil forms. I purchased two bases used for tuning coils, which had the coils removed from them. Handwriting in pencil on the forms matches that of Nikola Tesla. After reading my book, Los Alamos National Laboratory personnel went to one of the nearby salvage dealers and retrieved the contents which he had purchased. He apparently had taken everything except the forms which I had purchased because nobody else was interested in them, or had large enough trucks to carry them away. The old barn with all its recovered contents (except my two bases) is now preserved under lock and key for historical purposes.)

The weekend in 1979 following my viewing of the old wooden box at the Santa Fe Flea Market, I went to an Albuquerque salvage company, "King's Surplus City", on east Central Avenue, where I often bought military surplus things related to my research and inventing, and was pleased to find a very unusual pile of material stacked in a separate little pile. On top the pile was a device which I immediately recognized as a type of navigational "compass" which would be required on the early flying saucers I had witnessed during my back yard sleep-outs in Kermit, Texas, between 1949-53, which turned in increments or multiples of thirty degrees. The device

**showed thirty-degree increments, and had fourteen contacts—twelve for each of the directions, and two for the six-volt motor which rotated the compass ring to indicate the correct flight heading. The label showed it was manufactured on Oct. 1, 1943, in lot 127-178 (52 devices), for a project named "KT-p2". (I interpret this as "Kurskreisel-Tochterkompass - p2" which in English would mean "Navigational-Gyro Slave-Compass - p2". The device was called a "Peiltochterkompass" (a "polar slave compass"), which was controlled by signals from a "Meisterkurskreisel" (a "master navigational gyro"), which means that the device was half of an inertial guidance system, which requires a master gyro for its completion. The navigational master gyro, mounted on gimbals, was pointed at the Pole Star—either Polaris or Arcturus—and maintained the true north heading no matter where on earth the ship flew. Contacts on the gyro sent signals to the twelve contacts on the slave compass to tell it which of the twelve 30°-increments the ship was following. When the ship changed directions, a small D.C. electric motor came on and rotated the flight heading ring to the correct direction in relation to the Master Gyro.**

**Development of the systems was imperative for the flying saucer, since a magnetic compass would be worthless due to the "Faraday cage" effect of the electric field surrounding the saucer, which would block out the magnetic field of the earth. The master gyro was calibrated to the Pole Star prior to takeoff, and would continue to hold that position during flight, so that the "slave" ("-tochter" or "daughter") would have that as a reference, in order to actuate the six-volt motor and turn the geared compass heading ring to reflect the correct direction of flight, in 30-degree increments. This particular type of Peiltochterkompass was used on a circular ship which could turn peripherally—in 360 degrees, frontwards, sideways, backwards, etc., in increments or multiples of 30 degrees—while other devices were used on some of the latest types of German "V" (for "Vergehtungswaffen" or "Revenge-Weapons") rockets, which carried neutron bomb warheads, and were fired from underwater from special small Lafferenz Project submarines which were towed behind the latest "Electro-U-Boats" used in the final "Stalemate" which I detailed in the third edition of *Space Aliens From the Pentagon* (*Pentagon Aliens*). This submarine missile delivery system had been**

**tested (prior to the *Stalemate*) in the Baltic by Dr. Ernst Steinhoff, who died in Alamogordo, New Mexico in the 1990s.**

**This type of (now) relatively primitive navigation system was one which I knew had to have existed at the time of my 1946-47 night time sightings, in which the squadrons of high altitude flights from New Mexico turned in increments of 30 degrees. Similar systems were used in American military aircraft during WW II, but those systems were not designed to facilitate instantaneous sharp turns in 30-degree increments in the 360-degree range. Those devices were also not designed to be mounted horizontally as this one was, but were mounted on the vertical faces of instrument panels.**

**Tesla's ships were all of what I call the "linear" type, long and narrow. The Germans apparently contributed the circular or "saucer"-type ships, unless Tesla's designs which are presently concealed by government secrecy contain that type.**

**During my graduate school years in the department of art at the University of Texas, 1965-70, I often drew hypothetical diagrams of such a navigational compass.**

**Several years after I had published photographs of the KT-p2, there were web pages placed on the Internet by a German source---apparently our own CIA----which attempted to fabricate a different use for the devices, in order to counteract what I have said about it and to debunk me. The fatal error of this misinformation is the obvious omission of a "master" gyro which signaled the "slave" compass the true north heading by which it could navigate. Without the master compass the device would not only not be a slave compass, it would be inoperable. That omission shows the misinformation to be what it is, lies. Perhaps the agents who created the misinformation were not given access to those complete secrets because of "compartmentalization". It is however a fact that the devices, combined with other devices, were also adapted to other uses, such as the following into England of the VOR beams projected over England by the Allies to carry out night bombings over Germany.**

**The following photographs are of the 1943-vintage Nazi Kurskreisel-Tochterkompass—also labeled a *Peiltochterkompass*—which came from classified salvage at Sandia Base in Albuquerque:**

The Label

Peiltochterkompass
KT-p2

Gerät Nr. 127-178A-1
Anforderz FI-23374
Werk Nr. 10143
Hersteller: Gvy

**The letters "p2", which designated the 1936-38 project (two years when von Braun was in New Mexico before WW II) run by Tesla and von Braun, were incorporated into the official secret name-designation for the Nazi flying saucer project in 1943 Germany and displayed on the navigational Peiltochterkompass. "p2" is apparently a combination of rapidly-varying electrostatic potentials which initiate the propulsion of the ship and high-frequency currents which are used to aid in this according to Tesla theory. More on this later.**

**As explicitly detailed in Space Aliens From the Pentagon (repeated in Pentagon Aliens), in 1957, as an Air Force member of the 26th Strategic Reconnaissance Wing, SAC, Lockborne AFB, Ohio, with a Top Secret Clearance, I saw the original 16 mm reels of intelligence film made during WW II by my 8th Air Force Intelligence group in 1943, during the night-time bombing raids of**

the industrial center at Schweinfürth, Germany. That film was made by members of my same group during WW II, and the original film was in our film library. The B-17 bombers were "buzzed" by small, one-man type Tesla flying machines, which flew at very high speeds around the bombers, apparently in order to draw the fire of the gunners and interfere with their missions. This was a tactic apparently designed to distract the gunners so the "Komets"—Messerschmitt rocket planes—could get in close to use their 30 mm cannons on the bombers.

In the spring of 1945, as Russia closed in on Berlin, General George Patton took a special armored group deep into territory encircled by the Russians to the Nazi rocket research center at Peenemünde, on the Baltic. Attached to this group, disguised as "tank commanders", were Army intelligence officers specially tasked to retrieve all flying saucer materials and/or destroy the same at that project, along with retrieval of any technical drawings, etc., of strategic importance, before the area could be captured by Russian troops. On discovering 200 celestial guidance systems deep underground in a salt cave, General Patton was infuriated on being ordered to destroy the systems by his superiors, saying he could successfully get them back behind Allied lines, and being unaware of the fact that the technical drawings and specifications for the systems, as well as functional Nazi flying saucers were already in American possession. The documentation which I found of this trip to Peenemünde by Patton and his group has been removed from all war histories. The "fix" to hide the saucers was already on in 1945. And who are the authors of this fix?

Subsequent to my purchase of the Peiltochterkompass, security agents learned of my possession of it and knowledge of what it was, so they went to Mark King, son of the owner of King's Surplus City, and grilled him. He showed them a bill of sale from Sandia Base, pursuant to his purchase of the small pile of salvage from a Sandia Base employee who usually brought salvage from that base to sell to him. As it turned out, the man, who was under secrecy orders as an employee of that base, had stolen a pile of classified salvage and sold it for his own profit, using a "double-receipt" process intended to cover his tracks, but the security people at Sandia put the man under surveillance, caught him doing it again, and charged, convicted and

**jailed him.**

**They could not get the device back from me, because I was an "innocent purchaser", and because it would confirm the still-classified purpose of the device, in violation of security orders.**

**I had purchased an inertial guidance system from a 1943 Nazi flying saucer, which was subsequently brought to Sandia Base in 1945 under Operation paperclip, along with over 116 scientists and over 15,000 German military and technical personnel, brought to New Mexico as part of a secret 'deal', made prior to Operation Paperclip pursuant to a Stalemate, during which 24 German Electro-U-Boats, towing Lafferenz subs carrying rockets with neutron bomb warheads, in attacks concentrated on New York City. The first six of these boats in the first attack were bombed and destroyed. The remaining 18 in a second attack managed to coerce a secret conditional surrender for the German Third Reich.**

**This secret surrender was negotiated between President Truman and Admiral Karl Döenitz by radio telephone conducted through an intelligence Electro-U-Boat which was located off the coast of New Jersey and was revealed to be sunk there in a TV program (Nova or History Channel) in the late 1990s. Pursuant to that secret surrender, a provision was that the captains of the remaining 18 U-Boats went to the bottom with all hands on board. Consistent to this scenario evidence in that sub was that the captain had ordered an explosive charge detonated in the conning tower taking the sub to Davy Jones' Locker, presumably while the crew sang "Deutschland Über Alles". An underwater view of the conning tower showed that the explosive charge had been placed inside the conning tower as the surface was blown outward from it.**

**As to the question of what "p2" meant scientifically, a similar designation was used by Tesla to denote "potentials". I think it refers to the use of rapidly-varying electrostatic potentials in one direction to draw in the tubes of force to impart momentum to the ship and high frequency A.C. in the opposite direction to block the ether and stop the tubes from moving completely through the ship.**

**The drawing below is derived from information contained in Tesla's 1890s lectures and excerpted later in the part of this book regarding Tesla's "Dynamic Theory of Gravity", insofar as I am able to reconstruct it, since the actual theory was unpublished by Tesla.**

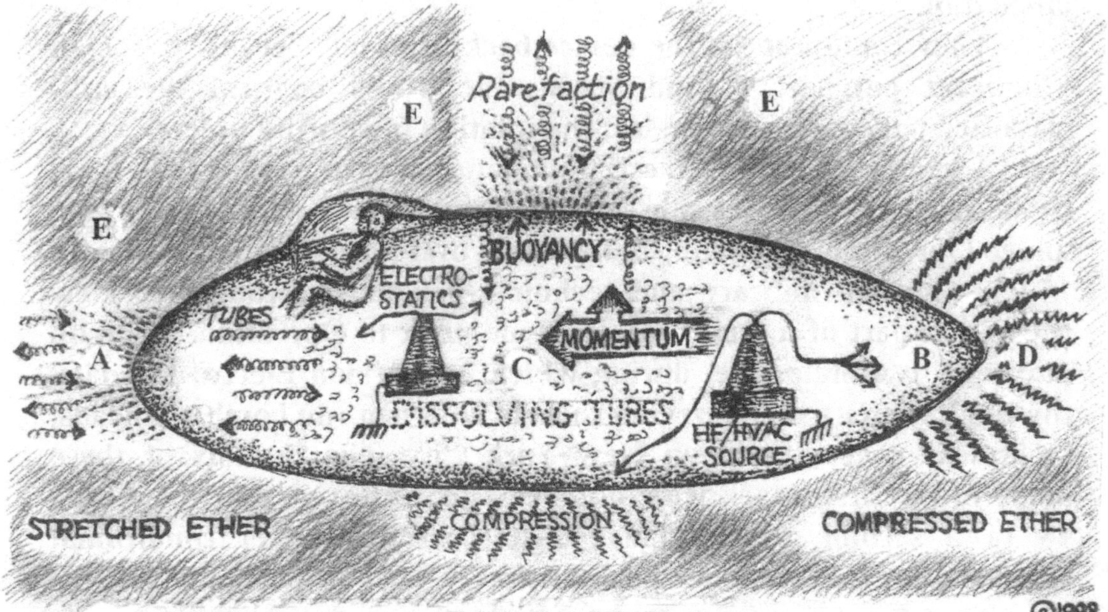


**ELECTROSTATIC PROPULSION - p2:**

A - Steady D.C. "Brush" Discharge (negative) - continuous exchange of Tubes of Force - ether rarified ("stretched")
B - High Frequency A.C. Potential - ether compressed - tubes blocked
C - Trapped Tubes of Force dissolved - momentum imparted at $2 \times 10^{39} \times fg$
D - Compressed Carriers - blocked tubes (positive A.C. corona)
E - Random Tubes of Force in free space

For horizontal motion or buoyancy, there must be, attached to the Tesla coil, emitter plates which create the sound waves in the ether with rapidly-varying electrostatic potentials. The frequency must not be as high as the microwave range, as the waves would not penetrate the shell. I see no reason why it is not possible for these emitter plates to be placed inside the outer shell of the ship, shooting the waves through the shell to the outside. These plates must have thick layers of high dielectric---such as barium titanate---on their backsides, so the waves are emitted only on one side in the direction desired. When a plate is flat, the waves are parallel; when convex, fan-shaped; and when concave, focused (such as for tractor-beams, to pull smaller objects to the ship).

A variable-concavity plate could focus on objects at various distances from the ship, as desired. This would be a mechanical problem.

This seems to explain some of the puzzling maneuvers of the

**ships observed by many. Meanwhile, a lower, rapidly-varying electrostatic potential may be placed on the entire ship's surface to "lubricate" it as it moves through the atmosphere without friction or sonic boom, while the directional waves are shot through this. This hypothesis appears to agree with what has been observed.**

**This technology has been highly developed for over seventy years since the early days at Los Alamos, years before the Manhatten Project came onto the scene.**

**I could only assemble parts of Tesla's technology like pieces of a jigsaw puzzle, from smaller statements he made at various times regarding it. As you can see in this drawing, based on Tesla's statements, there is some conflict between an ether which is supposedly "incompressible" versus the fact that Tesla sometime referred to "compression" of the ether by high frequency alternating current emissions in the atmosphere. I think he meant by this that the space between ether particles was eliminated. As you will see in the drawing, both the negative electrostatic 'brush' electrode and the A.C. 'compression' electrode would direct the ether carriers in the tubes of force through the mass of the ship from the negative electrode, and cause them to be dissolved in the conductors of the ship, imparting momentum to the ship, similar to what J. J. Thomson said. The electrode on the left would create a "veritable rope of air" so that, when paired with the opposite A.C. electrode on the right, would accelerate the ether carriers through the mass of the ship, dissolving them and propelling the ship. Tesla described this as a "pulling" action (not a "repulsive" action).**

**Many of Tesla's tests and experiments in Colorado were related to his propulsion system. A rapidly-varying electrostatic (D.C.) potential, according to Tesla, at one end of the ship creates sound waves in the ether which somehow draw in the momentum-carrying tubes of force. Meanwhile, a high frequency at the opposite end of the ship creates a positive potential according to Tesla, which causes the tubes of force at that end to be blocked, so that the tubes of force at the opposite end are accelerated into the mass of the ship and its contents, to be dissolved in the conductors to impart momentum to the ship. While the ship may be accelerated or turned so fast that a normal person inside might be crushed, the technology must operate in such a way that every atom of the ship is accelerated with the same**

**force, in the same direction, at the same time, so that no excessive force is experienced by a passenger. There was research in Germany during the war which discovered that a pilot in the prone position could withstand six times the G-forces as a pilot in the normal sitting position, but this could not explain the hairpin turns of the ship I witnessed in 1953.**

## Low and High Frequency Electrostatic Potentials And Tesla's Wireless Transmission of Energy

**Low frequency radio waves can pass through solid bodies, and tend to "hug the earth", while high frequency waves don't readily penetrate solid bodies, causing a "skin effect" on them, and don't "hug the earth", but tend rather to travel outward away from the earth in straight lines. In his Colorado Springs Experiments, Tesla was the first person to investigate high voltage waves of high and low frequency, and their terrestrial, atmospheric and solid-body effects. But remember, Tesla was working with a completely new type of wave, not electromagnetic ones but longitudinal pressure (sound) waves in the ether. The results of these experiments were so far-reaching, that whatever was said in current science journals about them had little effect, because they left other scientists so dumbfounded that they were paralyzed by mis-comprehension. This is not taught in electrical engineering schools.**

**It has been said that the basic difference between Heinrich Hertz' 'transverse' waves and Tesla's longitudinal waves is that Tesla's waves were actually electrostatic mechanical compression waves in the ether. Tesla actually said that what Hertz was observing was the same electrostatically-induced longitudinal pressure waves. In 1934, Tesla stated (Possibilities of Electrostatic Generators, Scientific American), speaking of the electrode on his wireless tower erected on Long Island in 1902, "It was to be charged to 30,000,000 volts by a simple device for supplying static electricity and power." Examination of the patent (#1,119,732, filed Jan. 18, 1902) shows "coil B" is a Tesla coil tuned to one-quarter wavelength, which produces rapidly-varying (D.C.) electrostatic potentials, resulting in sound waves in the ether, not electromagnetic waves.**

High frequency longitudinal waves also make it possible to project forceful mechanical waves to a specific, distant point. Tesla also developed a method with his special aluminum hemispherical bulb for sending waves or 'heavy currents' along a special "carrier beam" of U.V. or similar high frequency radiation which we don't know the exact nature of.

Powerful pressure (mechanical) long waves, carried directionally to a distant point, can in turn carry additional higher frequency waves superimposed on it, to deliver the high frequency exciting energy at the powerful long-wave peaks. These "sound waves" are apparently converted back into electrical energy by an appropriate Tesla receiver (or maybe just a conductor) at the target zone. This appears to have been the basis for Tesla's "Transmission of Electrical Energy Without Wires" invention, as well as his "energy beam" and "death ray" inventions, as the same system can be used for all three. The manipulation of these wave characteristics is inextricably involved in Tesla's propulsion. These inventions were undoubtedly stimulated by Tesla's study of 19th century "ether physicists", and his own brilliant intuitions when discovering the rotating magnetic field in Budapest, Hungary in 1882. This rotating magnetic field discovery was involved in his "brushless" electric motor because Tesla applied the rotatory force of magnetic fields related to the naturally occurring phenomenon spoken of by Faraday, Maxwell, both Thomsons, and several others known to Tesla, to his A.C. electric motor inventions.

The tubes of force had been discussed in 1881 in England by Thomson, who had no technology which could be used to apply the tubes to the propulsion of a ship, if that was what he had in mind, and never seemed to mention them in this respect until after Tesla had begun to move toward applying the discoveries to his intended "ideal flying machine" in experiments prior to his 1891 lecture at Columbia College in New York. But Tesla was influenced after he was apparently inspired by the writings of William Thomson and J. J. Thomson.

I think that Tesla discovered that by using a low frequency mechanical waves, he could cause, at the quarter wavelength, the eruption of a large pulse which could, by carrying a much higher exciting wave, excite the atmospheric gases in a region, to explosively

**expand them, followed by a tremendous implosion and the absorption of a great amount or heat in the area, causing immediate freezing. This process would create an initial explosive shock wave followed by the implosive return wave, similar to nuclear detonation. For example, by using a low frequency of say 50 cps, with a wavelength of 3,100 miles, a pulse could be made to erupt at one-quarter wavelength—775 miles, corrected to 751 miles by velocity factor. By superimposing an exciting frequency of say 50 megahertz or so, the atmospheric gases could be excited to cause a "cold implosion" which absorbs heat. By altering the wavelength and the direction of propagation, the results could be delivered to any terrestrial point. At least that is my interpretation.**

**The Tesla ionizing bulb, fundamental to some of Tesla's "beams", is a solid aluminum hemisphere, enclosed in a glass vacuum envelope, excited by high voltage D.C., applied to the center of the hemispherical end. A polarized beam is emitted normal to the flat face center. The frequency and force is determined by the voltage. The ion beam polarizes, concentrates, and guides a long wave amplified by Tesla's "extra coil", to a distant target. This amplified higher frequency wave superimposed upon it will "ring" in the target area at one-quarter wavelength, in fact ring many times.**

**Outside the bulb, in close proximity to the beam, the rapidly-varying high voltage electrostatic impulses create the pressure waves which are emitted by a ring-shaped electrode, and is directed along the beam's atmospherically created conductive path, projected toward the distant target where the wave peak delivers it at its "point" of greatest potential.**

**The June 30, 1908 Siberian explosion/implosion at Tungushka may have been a risky test or inadvertent result of Tesla's giant Wardenclyffe oscillator, and may have been an undisclosed reason as to why the project was discontinued, and eventually the tower was torn down in 1917. It should also be pointed out that this method involved the absorption of heat energy "from the environment". This was yet another great heat energy-harvesting discovery. Tesla's method "opened up" the atmosphere to this possibility, consistent with what he said about nuclear energy coming from "the environment". And maybe that is what Leland Anderson was referring to when he delivered Tesla's eulogy at his 1943 funeral,**

praising Tesla's ability to "...pluck heat units out of the thin air."

The Tungushka event may have also been the impetus for a "very tempting offer" made to Tesla by V.I. Lenin, mentioned in retrospect in a November 29, 1934 letter from Tesla to J. P. Morgan, Jr., (microfilm, Library of Congress). The Lenin offer would have to have been between 1917 and 1934, and was mentioned unsuccessfully by Tesla in attempt to tantalize Morgan.

The following illustration was not a Tesla one as thought by some, but turned out to be the idea of what another inventor apparently thought was Tesla's idea, which he himself was trying to cash in on, since Tesla never revealed it in detail. Tesla had something better, a particle beam which was finer in cross section than a hair, of tremendous power. The government has never revealed the details of Tesla's so-called "Death Ray" invention, but it is hinted at by Tesla's announcement of Sept. 22, 1940 in the New York Times, in an article by William L. Laurence, of Tesla's *Teleforce* discoveries. It should be noticed that the waves here are focused.

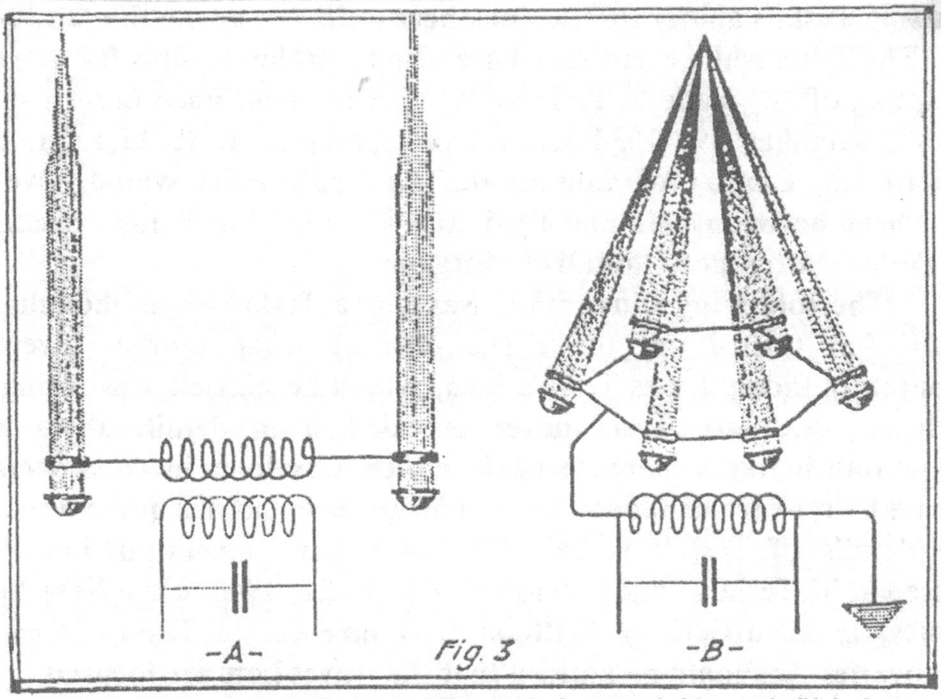


This illustration, from a July 11, 1934 *New York Times* article, shows several of the special solid aluminum Tesla bulbs which, as Tesla stated in 1940 (see page 27), no longer required vacuum. The current is placed on the beams by means of the ring-shaped electrodes near each bulb, fed with currents from the coils.

## Tesla's TELEFORCE Discoveries

**Tesla's Teleforce discoveries, evolved between 1900 and 1940, supposedly came in four parts, consisting of a "new" way to produce rays in free air without a vacuum, a "new" way to create a "very great electrical force", a "new" way of "magnifying" the force, and a "new" means of space propulsion[2] . At least that was what was said in books. When I consulted the actual article, I found quite a difference. What Tesla called "His Greatest Achievement", was "an apparatus by which mechanical energy can be transmitted to any part of the terrestrial globe" and which had four practical possibilities (aside from the energy transmission part), among which were a means of unfailing communication, a means for scientists to lay bare the physical conditions of the earth, and enabling them to determine all the earth's physical constants. He called this "tele-geodynamics", the determination of earth forces at a distance. He had an apparatus**

2[2] William L. Laurence, *New York Times*, Sept. 22, 1940, 11, p. 7.

which he said was "ideally simple" consisting of a stationary part and a cylinder of fine steel "floating" in air. By impressing powerful impulses on the floating part, he said, it could be caused to react with the stationary part which then would transmit mechanical energy through (the ether in) the earth. This involved a "new amplifier for a known type of energy" which could produce the impulses through the earth so they can be picked up whenever and wherever needed.

I have discovered that the "stationary part" is his "Coil For Electromagnets" (Patent #512340, filed July 7, 1893) which was his bifilar coil which stores 250,000 times the energy of a ordinarily wound coil of the same size and number of turns. It is this coil which stores so much energy that it causes the steel cylinder to "float". The "known type of energy" was of course the magnetic field. Just exactly how he subjected the steel cylinder to powerful impulses is another of Tesla's discoveries, which was at least 35 years old at that time. It was probably his single terminal ("pancake") coil. He said it "will be considered absolutely impossible by any competent electrical engineer", a "new method and apparatus for producing direct current without a commutator", considered impossible since the days of Faraday. These "new" discoveries—all of which could work together—were actually a juxtaposition of and different application of Tesla's devices already tested separately in Colorado Springs experiments or later in New York or Wardenclyffe, with the exception that by 1912, he had tested a missile powered by wirelessly projected "mechanical" energy, and he had also tested a small, one-man flying machine prototype in which he apparently ventured out his hotel window at night over New York City in experimental tests flights, which served as an experimental basis for a much larger and more ambitious set of space ships which he had designed, which were repeatedly rejected by the U.S. government, probably because they thought he was nuts. In one interpretation, Tesla's teleforce discovery was apparently a "tractor beam" technology which could reach out and lock onto many tons of air molecules—or onto the incompressible ether—by which means the "very great electrical force" could be used to pull against to propel the ship up or forward or backward without air or inertial resistance.

**Tesla said of his flying machine in 1911[3], in pertinent part, "Its lifting power will not depend upon any such delicate devices as the bird has to employ, but upon positive mechanical action." By "positive mechanical action", Tesla was referring to his "sound waves in the ether" which travel at the speed of light, which are mechanical in nature. Tesla had stated that his training was in mechanical engineering, and that he was primarily a mechanical engineer. In a 1929 article in the New York Herald Tribune[4], Tesla stated** ***"I satisfied myself that what he (Heinrich Hertz) had observed was nothing else but effects of longitudinal waves in a gaseous medium, that is to say, waves, propagated by alternate compression and expansion. He had observed waves in the ether much of the nature of sound waves in the air." And, "The velocity of any sound wave depends on a certain ratio between elasticity and density, and for this ether or universal gas the ratio is 800,000,000,000 times greater than for air. This means that the velocity of the sound waves propagated through the ether is about 300,000 times greater than that of the sound waves in air, which travel at approximately 1,085 feet a second. Consequently the speed in ether is 900,000 x 1,085 feet, or 186,000 miles, and that is the speed of light."***

**Even the Nazis were skeptical that the Tesla propulsion technology might work, being unconvinced until Wernher von Braun had actually tested the concepts at Los Alamos in 1936 and 1937, and obtained the final approval of Hitler. The large conical coil form, constructed of rough-sawn pine boards held together with dowels and shellacked, which I saw there from the old barn was about ten feet from top to bottom, a couple of feet across at the top, and about ten feet across at its base. I could see where heavy wire—then removed—had been wound on the boards from which the form was constructed. Almost immediately after von Braun's return to Germany in late 1937 in time for Christmas, construction was begun on large "cigar-shaped" ships (similar to those shown in the illustrations Tesla commissioned in 1914), using at first the iron pressure hulls of U-Boats, which were already equipped with a hermetically sealable system carrying diesel generators for use in the**

---

3[3.] *Tesla's New Monarch of Machines, Oct. 15, 1911*

4[4.] *Nikola Tesla Tells of New Radio Theories*, New York Herald Tribune, Sept. 22, 1929

**water or the atmosphere, and possibly the helium-based "nuclear" generators which I had heard about in 1947. With the huge difference between the weak interaction of gravity and the much stronger interaction of the electrical attractive force, "Weight is no object" Tesla said.**

**Since Tesla had actually received offers from Russia prior to that time, I had considered the probability that he had cooperated with von Braun in the Los Alamos experiments, through some of the unidentified three laboratories he said he was working with in 1937. But I already had the information that von Braun had been involved in a 1930s Los Alamos project from my friend Peter van Dresser. The old box from the p2 project which I saw in 1979 filled in the blanks. In 2008, I found evidence that indeed Tesla was here in New Mexico in 1936, and there was really not very much else he could have been doing here at that time. In addition to the work at the p2 project Tesla also was involved in 1936 in the testing of his wireless tranmission of energy system to run a train in Santa Fe, New Mexico, another activity which there is virtually no information on, even in Santa Fe. The spooks did a good job of cleaning out all references.**

**In looking for a probable Nazi spy, there was one who had wormed his way into Tesla's most intimate confidences, who was later convicted and imprisoned for his espionage activities in America, but the exact nature of his activities, other than the dissemination of Nazi propaganda, was never specified in the public records. This author's belief is that Tesla's sale of technology to German Admiral Alfred von Tirpitz in 1914, following rejection of this technology by the U.S. government, included the technology for Tesla's electric flying machine, not developed until 1936-38. Tesla's connection with Georg Silvester Viereck was the most probable explanation for how Tesla's propulsion discoveries attracted the attention of the Germans and found their way to the internationally funded "p2" project (funded by the Germans, the National Geographic, the Simon and Florence Guggenheim Foundation and the Smithsonian Foundation).**

**In his 1935 birthday announcement[5], Tesla was extravagant,**

---

5[5]*New York Times, July 11, 1935 (supra)*

treating about 30 reporters, press, camera men, news and sound reel representatives to a lavish gourmet luncheon in a private dining room at the Hotel New Yorker where he had been living for the previous two years. This indicated a new source of revenue for Tesla, most likely from his German connections for the p2 project.

The fact that von Braun's pre-war New Mexico adventures have been subsequently concealed from the public by the Secret Government—a degree of secrecy which cannot be explained except by reference to the Nazi flying saucer project along with appropriating my friend Peter van Dresser's and his four colleagues' rocket engine design—is compelling.

The fact that I was a personal friend to an American rocket scientist named Peter van Dresser, who was angered by Dr. Robert Goddard's pre-war work with Wernher von Braun, and who was present in New Mexico in the '30s and was personally familiar with von Braun's secret Los Alamos activities left little remaining to be proven, since there is no longer any security basis for concealing such events, if "rocket research" was the only thing von Braun's pre-war American presence involved. He had already been appointed, just prior to his trip to America, as the Peenemünde research and the national German rocket research leader, and had been promoted by Hitler and Heinrich Himmler to Uberstürmbannfuehrer in the SS, and was given a "leave of absence" by Hitler to make the venture to America. After all, he was at that time in the Wehrmacht as well as the SS.

Willi Ley, the real rocket genius of Germany and von Braun's tutor, was the first scientist invited to America by Peter and the New York branch of the American Rocket Society in 1934. In 1937, Ley was staying at the Greenwich Village, NYC apartment of Florence and Peter van Dresser, applied for American citizenship papers that year, and married a Russian ballerina named Olga Feldman, so von Braun was apparently invited by Dr. Robert Goddard, by his arrangements with the Germans, to fill the empty shoes of Ley. I suspect that all those promotions of von Braun after Ley's defection were to insure that von Braun did not also defect and become an American citizen.

**As shown by archival proof [6], van Dresser's group was the actual inventor of the first successful liquid-fueled rocket engine—called "A Dependable and Efficient Regeneratively Cooled Liquid-Fueled Rocket Engine" by Peter—the particular one that Dr. Robert Goddard "appropriated", and turned over to Wernher von Braun, who used it on his V-2s. Peter was co-creator of the American Rocket Society and was the editor of its publication, *Astronautics*. But when the military became involved in the research at Mescalero Peter left because he was a Quaker pacifist and took Willi Ley with him so von Braun was brought in by Goddard from the Third Reich as a replacement in 1936.**

**One of the most telling pieces of evidence regarding the intentional misinterpretation of Tesla's discoveries—especially just after Tesla's death in Jan., 1943, at which point all Tesla's papers and apparatus fell into the hands of the FBI—is the obvious meaning of the name for his system—TELEFORCE"—which means "force at a distance". This must be seen in the context of the four listed parts of the system to be used together, being clearly applicable to flying saucer propulsion, once I pointed out the fraud, and correctly interpreted the statement in my 1993 publication.**

**I have to hand it to the spooks, for doing such a good job of concealing the connection between the flying saucer and the only man who could have invented it—and whom the public should most readily accept as the inventor who rightfully deserved the credit for it —even though he never had a fair opportunity of giving his greatest discovery to the world as an American, for lack of opportunity to build a working model of sufficiently developed potential to impress us. And the German-financed p2 Projekt gave him that opportunity after the U. S. government spit in his face. The reason why Tesla never had the opportunity was because of the secret "bill of attainder" placed on his name by the greedy monopolists who were and still are afraid of technological freedom and a truly free market of ideas, goods, and services. Tesla was a pariah and no one with money and power could touch him in America without being incinerated by the lightening strikes from the money gods in New York.**

---

6[6] New Mexico State Archives, Historical Division, Peter van Dresser Collection.

# CHAPTER II: EXPLORING THE NATURE OF THE ETHER

## The Early Background

In 1997, after many years of research and thought regarding the qualities of the ether, I decided to see what others had actually said, before completing this book. Some of the material available to me was second-hand information, being what someone said someone else had said. I wanted to determine whether or not these "repeated statements" were accurate, and if not, find the correct ones. After some difficulty, I accessed a copy of the classic, *A History of the Theories of the Aether and Electricity,* by Sir Edmund Whittaker (1951 revised and enlarged second edition, Vols. I and II). The 1910 first edition was probably the most comprehensive book ever published on ether theory as it existed in 1910, except it omitted the ether theory of Tesla. The first edition was printed in Scotland, under the title *A History of the Theories of the Ether and Electricity, from the age of Descartes to the close of the nineteenth century.* The second edition incorporated "new material" related to quantum mechanics. Classical theory was in Vol. I, while Vols. II and III gave the origins of relativity and quantum theory up to 1926. Neither the 1910 first edition nor the 1951 revised edition mentioned Tesla at all, but the chronological sequence of attempts at electro-propulsion by the Europeans appeared to be "skewered", with little or no attempt in either edition to point out what the scientists were actually aiming at in terms of electric propulsion, which was clearly their goal, or to point out this goal as a topic. Though information regarding Tesla's work was omitted, there is much valuable related information in the historical sense in terms of many experiments.

In his preface, Whittaker stated that where space is not considered "emptiness"—except for its property of *propagating electromagnetic waves* (a concept which is completely wrong according to Tesla's discoveries regarding the pressure waves)—the development of quantum electrodynamics referred to a vacuum with a new status as the location of 'zero point' oscillations of the electromagnetic fields and zero point fluctuations of electric charge and current, as well as a polarization conforming to that of a dielectric constant, which is non-unity (i.e., which contains energy). This implies awareness of Tesla's Primary Solar Rays, but quantum theory inherently excludes the possibility of "zero-point" fluctuations, since all such waves are always regarded as possessing one-half "quantum", and this is an internal contradiction of the theory. And the probability remains that all this is completely wrong.

Whittaker said that an alleged failure by ether theory proponents to observe the earth's motion relative to the ether, caused the downfall of the theory, but the facts show that the relativists, backed by corporate publishing interests, hastily dug the grave and threw the dirt over the corpse of ether theory while it was still alive, for the wrong reasons (or maybe the "right" reasons considering the lies propagated by powerful interests). Much of the "error" was due to focusing on the weakest proponents who failed to realize that a portion of the ether is "carried along" by the earth, and that an optical aberration—an "ether drift"—would have to be observed at some point away from the earth, in order for an ether to exist. It was asserted that the failure of this observation on the earth was the "death" of "the" ether theory. By advertising ether theory's weakest and most erroneous proponents and ideas, the media made the RQM's hasty funeral of ether theory easy, and paved the way for the gullible public's acceptance of relativism. But remember, truth can flow from lies, but not lies from the truth, as stated by Schopenhauer.

I concluded that Whittaker's 1951 edition intentionally skewered important facts which already should have been known to the author concerning Tesla's theory, which I have attempted to straighten out for you in this book. I now feel more comfortable with my interpretation of Tesla's theory, than I did prior to reading Whittaker's book, but that was only because I knew what I was looking for. The facts I found are almost exactly as I said in my second edition of *Space Aliens From the Pentagon*. It was unthinkable that a 1910 or 1951 book on the ether failed to mention Tesla or his ether theory, which contains distinct features not found together in any other ether theory.

By gathering and reorganizing Whittaker's material on electro-propulsion into more proper order, I discovered much "new" ("old") documentation, that 19th century scientists were in pursuit of electro-propulsion, and that there were interesting similarities

**between the thinking of certain scientists of the era and my suspicions, which could eliminate some erroneous concepts about what they actually meant, and show similarities to what I had observed on numerous occasions in the behavior of what were obviously Tesla's "flying saucers", since the "proof of the pudding" is when it flies across the dinner table or the sky. It was my intent to examine known theories leading up to Tesla, to see what he contributed, and to determine what happened since that time to obscure, misinterpret and ultimately conceal the truth. To this end, I present for you a brief history and documentation of pertinent ether physics.**

## J. J. Thomson's "Electromagnetic Momentum"

**Returning to the concept of the "carry-along" ether, Max Planck, originator of quantum mechanics, followed this form of ether theory in part, saying that the ether had to be irrotational, carried along at the same velocity as earth near its surface, and compressible and subject to gravity like the atmosphere, in order for the theory to be true. At best, Planck's beliefs showed a weak mechanical conceptualization, something unexpected from one purportedly so adept at "mechanics".**

**J. J. Thomson (1856-1942), discoverer of the electron, expressed a different and more plausible ether theory. Thomson had theorized —based on Maxwell's earlier theory—that a charged spherical conductor moving in a straight line must produce displacement currents in the medium[7]. But this did not purport to be electro-propulsion. In 1893, twelve years later, he claimed to have discovered "Electromagnetic Momentum", saying "...in an electromagnetic field there is stored in every unit volume an amount of mechanical momentum, proportional to the vector product of the electric and magnetic vectors".[8] Thomson also was said to have developed the theory of moving tubes of force—which harked back to Faraday's earlier work—saying in 1891 that molecular structure is closely connected with tubes of electrostatic force, with magnetism regarded**

---

7[7] Phil. Mag. xi (1881), p. 229.

8[8] J. J. Thomson, Recent Researches in Elect. And Mag. xxxi (1893), p. 13.

**as a secondary force. He stated that "...the aether itself is the vehicle of mechanical momentum, of amount (1/4πC) [D·B] per unit volume."[9]**

## Tesla Shows Up

**It is interesting to note that these 1893 "discoveries" by Thomson, although following his initial speculations based on Maxwell's theory, also followed Tesla's lecture before the American Institute of Electrical Engineers at Columbia College, N.Y., on May 20, 1891, in which he described earlier experiments in which he had pre-empted Thomson's work with experimental verification, together with a demonstration of techniques and machinery built by him which obtained frequencies and voltages which W. Thomson (Lord Kelvin) had said earlier in England, were "impossible to achieve". In that sense, the "good old boys" in Britain should have admitted that Tesla's work was prior, after having admitted that they were unable to create the equipment necessary to produce the required frequencies to investigate and to experimentally verify these phenomena. But Tesla's theory was different, based on electrostatics not electromagnetics.**

**J. J. Thomson's theory, which supposedly linked electromagnetism with momentum in a mathematical and unified way, on paper, somewhat repeated Tesla's earlier lecture of 1891, proven by his earlier experiments, but Tesla's theory was different from Thomson's and Maxwell's. Thomson's electromagnetic momentum could be theoretically created only with Tesla's contrivances, which existed before Thomson's "discovery", and there was no way that Thomson could have reduced his theory to practice or experimentally verified it, because it would not work since electromagnetic waves cannot be transmitted through the free ether because it is not a solid.**

**Though the facts show that Thomson was prior to Tesla in attempting to publish a valid theory, Tesla could not publish prior to perfection of his invention, which was not a factor affecting Thomson. Tesla was apparently the only one to prove his theory by**

---

9[9] Phil. Mag. xxxi (1891), p. 149; Recent Researches in Elect. And Mag. (1893), chap. 1.

**experimental results. Tesla deferred often to Thomson, even though Thomson attempted later to claim Tesla's alternating current and high frequency discoveries. Thomson's earlier work[10], concerning a charged spherical conductor moving in a straight line (but involving no "rapidly-varying electrostatic potentials") had undoubtedly commanded Tesla's respect and probably influenced his work, but Tesla's 1884 discovery of the rotating magnetic field had beaten them all to the punch, and documentation shows that because of this he had arguments with his "learned professors" in Graz, who said that he must be "crazy" for having such ideas during his 1870s student days to prove it.**

**It must have worked a horrible hardship on Tesla, to remain silent about his *Dynamic Theory of Gravity*, which he did not publish for fear of his technology falling into the wrong hands. It was like the Wardenclyffe tower, turning in the wind of incompleteness for lack of money, created by having been falsely induced into building an expensive system which would exhaust his energy and financial resources to the point of bankruptcy.**

**In respect to momentum, I would analogize it as follows: When a body is held by a spinning or arcing arm, it contains the changing momentum created by angular movement. It must continually shed some "old" momentum while accumulating "new" momentum in the new direction. When released, it moves in a straight line vector corresponding to the state of the tubes of force which have imparted their momentum at the time of release. While held by the arm, centrifugal force was created by the tendency of the tubes to impart momentum in a straight line, which created the resistance to change by the tubes at any moment. Once released, "something" (the tubes of force as they then existed) determined the body's direction of movement and velocity. These facts are consistent with Tesla's own concepts, in contrast to other theories of the ether or tubes of force, as I will explain later in greater detail.**

**Most of Thomson's 1881 work harked back to Maxwell, and before him, Faraday, who had stated in 1831 that movement of an electrostatically charged body is equivalent to a current, which was**

---

10[10] Phil. Mag. xi (1881), p. 229 (Supra.)

**described as a tube of force created by movement and momentum[11].**

**In his consideration of the nature of "ponderable matter", Faraday had suggested that an atom may be nothing more than a field of force—electric, magnetic and gravitational—surrounded by a point-centre, which is completely penetrable[12]. This statement shows how Faraday's opinions had temporarily been swayed by the relativism of Newton and Boscovitch. Fresnel later stated, 'Upon considering the aberration of the stars, I am disposed to believe that the luminiferous aether pervades the substance of all material bodies with little or no resistance, as freely perhaps as the wind passes through a grove of trees'. Fresnel's concept presupposed that the ether surrounding the earth was unaffected by earth's motion, and generally agrees with what is actually observed, except for his failure to consider how the earth's dynamic electric field affects the ether, and how the ether is not in motion relative to bodies at relative rest on the surface of the earth or within its electrostatic field.**

**Along with his ca. 1831 theory of an (elastic) solid ether, and lines of magnetic force, Faraday considered all space as "filled".**

**Wm. Thomson (Lord Kelvin, 1824-1907), at age 17, based on Faraday's earlier announcements, introduced into mathematical science the idea of electric action at a distance by means of a continuous ether. In 1846, he analogized electric phenomena with mechanical elasticity. It was his early concept that material atoms move through the ether 'without displacing it or compressing it'[13], a concept later to be adopted partially by Tesla in saying that the ether 'behaves as a fluid to solid bodies, and as a solid to light and heat'.**

**About 1853, Bernhard Riemann (1826-66) conceived an ether which resisted compression and changes of orientation[14]. Riemann thought the resistance to compression caused gravitational and electrostatic effects, while the resistance to changes in orientation caused optical and magnetic phenomena, but he failed to develop these concepts further. These early Riemann insights were however very influential on later thinkers. He conceived a stationary ether**

---

11[11.] Faraday, Experimental Res. Sec. 1664; Maxwell, Treatise, Sec. 768-70.

12[12.] Bence Jones' Life of Faraday, ii, p. 178.

13[13.] *Letter,* FitzGerald to Heaviside, 4 Feb. 1889.

14[11.] Ann. D. Phys. cxii (1867), p. 237; Riemann's Werke, 2e Aufl., p. 288; Phil. Mag. xxxiv (1867), p. 368.

based on the generally accepted assumption that, as the earth moved through space, the ability of one to see a star with a steady image disproved a dynamic ether. As to the propagation of electric action through space, Riemann proposed a new equation in which voltage changes outward from charges with a velocity C, but he also failed to follow up with a proper consideration of the properties of the ether, because he died too damned young. Some of Riemann's insights were verified almost 50 years later, ten years longer than his entire lifetime.

Years later, Maxwell, after reading Faraday's 1831 *Experimental Researches in Electricity* (3 Vols.), formed a mechanical conception of the electromagnetic field, connecting Faraday's magnetic lines of force and electrical ideas with the mathematical analogies of Wm. Thomson, carrying Faraday's concepts a step further (I have always like to say that Maxwell "put the math to Faraday", because Faraday couldn't do much math, though he is recognized as the greatest "experimental philosopher" in history).

Faraday was a former Irish bookbinder's journeyman who got his early education from the books he was binding, and who at age 21 happened to write a letter to chemistry professor Sir Humphrey Davy in 1912 and so impressed him that the following spring he was appointed a chair at the Royal Institution. Faraday had remarkable insights right from the beginning, which remain long after his death, with so many observations and discoveries which were made without resort to tiresome pages of math, lucky for Maxwell, who made his own contributions in the process, although many were wrong according to Tesla, when it came to the ether and electromagnetic waves in free space.

Faraday had erroneously stated that induced magnetism in iron exists only within the iron, was zero in external free ether, and was the basis for electrical induction. Maxwell's "electric displacement in a dielectric" was analogous to the magnetic induction B, which he said may have a value different from zero even in the free ether. According to Maxwell, there is a displacement (i.e., an electric movement away from equilibrium position) wherever there is electric force, whether "material bodies" are present or not (thus carrying his

**electric displacement into the ether)[15].**

**It had occurred to me that since the earth and bodies "at rest" on it are in uniform motion (or uniformly changing motion) in respect to an ether which is relatively stationary because "carried along", that there were differences in electric displacement of ether in the space within solid bodies and external to them in space within the free ether, due to the differences in dielectric constant between the two. I deduced from this that the constant movement of bodies through a stationary ether-filled space, carrying their charges with them, create displacement currents.**

**Maxwell had represented Faraday's lines of magnetic force (vectors of magnetic induction) as the velocity of an incompressible fluid, based on Faraday's suggestion years earlier that there may be a "dynamical condition" similar to electric current, and that the 'physical lines of magnetic force are currents' of this "fluid". This agrees with Faraday's "dynamical condition" of the electric field 'carried along' by the earth, equivalent to a current, which appears to me to possibly produce the earth's magnetic field and gravitation, by effects on the stationary ether within earth's dynamic electric field. It also implies that magnetic lines of force are the "equal and opposite" reaction to electrical current as Faraday stated, and vice-versa, though magnetic fields are more the secondary effect of electric currents as Tesla said. The magnetic field is equally important however, in its creation of the rotatory motion imparted to electricity, which is responsible for the "tubes of force", which carry momentum which maintains the perpetual motion of celestial bodies in the universe, and guides electric currents through the ether forever. And this sounds very much like it is in agreement with Tesla's Dynamic Theory of Gravity as well.**

## Cosmic Motion and Local Momentum

**The earth orbits the sun at approximately 18.5 miles per second (ca. 66,600 mph). The solar system is orbiting the center of the Milky Way much faster, and the Milky Way orbits the Magellanic Clouds much faster, and so fourth. This hierarchy of motion and velocity**

---

15[15]. Maxwell, Scientific Papers I, p. 451; p. 526.

**increases from the moon, to the earth, the Milky Way, the Magellanic Clouds, etc., although these pale into insignificance relative to electromagnetic velocity, since movement of force is time-relative. In all, the earth's velocity and polarity relative to the ether continually varies according to its rotation, orbital position, and galactic cycle. When between the sun and the Milky Way's center, for example, it moves relatively slower than when to the "outside". As the earth rotates each day, its polarity of movement relative to the sun rotates 15.2 degrees per hour, with a rotational surface velocity at the equator of approximately 1,041 mph. This continuous change in polarity is so gradual that it is sensually imperceptible, but probably contributes to modulating the gravity effects (and may even influence our emotional states!).**

**According to W. B. Morton[16], an increase in charge of a body increases its "virtual mass", which increases the work required to move it by amount $2e^2/3ac^2$. This was in correction of J. J. Thomson's figures. Such a case suggests a dynamo consisting of large, insulated conductive balls on each end of a shaft turning at its center on a perpendicular axle, the balls being alternately charged and discharged to increase and decrease their "virtual" masses, thus rotating the shaft and turning a generator. But this possibility appears to be a fantasy, since the "increase in work required to move the ball" (in Thomson's theory) may be related to an increase in the resistance to the penetration by the ether, caused by the presence of the extra charges, which according to Faraday created a "Faraday cage" effect which shuts out the magnetic lines of force. Without the magnetic lines of force—or with suppressed, deflected or diminished lines of force—the rotation of the tubes of force by the magnetic field may be diminished or stopped, resulting in a resistance to increased momentum. Such propositions gives ample food for thought.**

**If Tesla had continued to increase the earth's electrostatic charge with his gadgetry, he might have thrown the earth's timing off, slowed it down, and possibly caused it to drift toward the sun or away from it. But that is what he said was possible. Being aware of these facts he later addressed them and even listed them as future technical possibilities, which also corroborated his developments in**

---

16 W. B. Morton, Phil. Mag. Xli (1896), p. 488.

**the field of propulsion, in case anybody noticed.**

## Local Momentum and Celestial Movement

**It's an odd thing, momentum. If a body is in motion relative to earth, its momentum is in the direction of motion, while as a body at rest on a spot on earth, it still has momentum along with the spot's ultimate motion relative to the universe. The motion is compound, the ultimate momentum of a body being truly a vector which represents all the forces acting upon it. Yet as a body at relative rest on the earth, it appears to defy all the other vector quantities because in it's stationary position it becomes a virtual part of the earth.**

**The earth rotates at over 1,041 mph at the equator, so there is that momentum, but the earth's orbital speed and path around the sun means that the earth's more general path in the solar system is in the direction of its orbit. A spot on the earth's rotating surface would appear to a distant observer to be moving slower in its direction of orbital motion over a period of half a day, yet it is always moving in its orbital path at over 66,600 mph. The rotation alternately diminishes and increases the speed of that "spot" through the universe, gradually rotating its ultimate solar orientation once every 24 hours. With this compound movement and momentum—of the earth, the solar system, the Milky Way, etc.—we have a more complex pattern of movement, and since a body is in constant, compound, gradual, ever-changing motion relative to the ether and the universe, some interesting effects are bound to be discovered. These effects are not, as described by Newton's laws, due to an attractive force acting at a distance, through an empty space. "Force at a distance" requires an ether relative to Tesla's Teleforce discoveries and his *Dynamic Theory of Gravity*.**

**Due to the relatively gradual orbiting and rotating changes in respect to earth's motion relative to the background ether, a relatively small amount of "externally acting" force ("gravity") would have to be ever-present in order for Newton's laws to appear to apply even on the purely descriptive level, except for the local ether phenomenon which I will specify in further detail, which insures "force-free" perpetual motion, despite the gravitational effects of the ether. While I agree with that part of Einstein's theory that says gravity is not due**

**to the direct action of one object on another (no "force at a distance"), I disagree with the part that says it is due to the idea that a large spinning body "wraps" space around itself and "curves it". Without an ether there is nothing to curve, and with it, there is no Relativity Theory and "no curved space" (as Tesla said).**

**In early 1995, in exchange for a copy of *Space Aliens From the Pentagon*, I received a copy of Eric Lerner's *The Big Bang Never Happened*, and was pleased to find that there are some in the academic world of physics (or astrophysics), who share at least in part, my own views on the 'Big Bang Theory', although they don't go so far as I do in saying the theory has been promoted by the coercive energy monopolists, along with Relativism, because of its apparent negation of the free energy available from the universe via the ether. This universal free energy is due to what I call the "Law of Conservation of Perpetual Motion". I found Lerner's book refreshing, not to mention useful in charting some of my own if not Tesla's ideas on a more cosmic scale.**

**Lerner's relatively unsung hero is Hannes Alfvén, a Swedish astrophysicist and cosmologist, whose earlier "academic burial" by the Relativists was somewhat reversed in 1970 by a Nobel Prize awarded to him for his development of cosmic magneto-hydrodynamics and his concept of frozen lines of magnetic force, a theory later modulated in part by him. Alfvén had in a way replaced the word "ether" with "plasma", and strongly disagreed with the Relativist's misuse of mathematical theory, which he said 'must always be the servant of physical understanding and close observation, rather than its master'. He claimed that giant currents through space, from the sun through the planets, along frozen lines of force, actually transfer angular momentum to the planets. This discovery, together with his work on cosmic MHD events, places Alfvén's work very close to Tesla's Dynamic Theory of Gravity.**

**The appearance of such books as Lerner's is a positive development, which represents dissent which seems intolerable to the Relativists. The Big Bang Theory is in decline among scientists, despite all the NASA false propaganda in the press and mass media attempting to perpetuate and support this false theory. I recommend Lerner's book, even though he has not yet evoked an affirmative**

**support for ether theory, but has dared to challenge Relativism and the Big Bang cosmology.**

**The actual course that a particular spot on earth takes through the universe is naturally more complex than that of the sun, since the earth and the spot on it represents a smaller cycle within the larger. Relative to the background ether, the entire interrelated movements of the solar system and Milky Way takes our spot in a generally wavy path, along which our velocity relative to external bodies periodically accelerates and decelerates.**

**The same principles of momentum which apply to the cosmos must also apply to the motion of immediately local bodies, except that the effects are skewered by the presence of earth's pesky electric, magnetic, and so-called "gravity" fields. What Alfvén had touched upon in his long-unrecognized discovery of the giant electric and magnetic currents through space, are similar to unrecognized forces operative on a local scale as well. As our planet hurls through space at ca. 70,000 mph, and us with it, there are things happening on a smaller, even microscopic scale, which play important parts in this giant panorama.**

**The big question—the one which those who promoted the Big Bang cosmology sought prematurely to put to rest—is how can everything stay in perpetual motion, while under the Relativist theory the law of conservation of energy is violated—and thereby the Relativist theory also refuted? The still-born theories of Tesla were forever gnawing at the monopolists' throats, along with those of Alfvén, so they wanted to obscure all truly competing theories, and to enshrine the Relativists and Big Bangers in the annals of science for all time. Their method was to fabricate a scenario in which one original giant explosion was the 'origin' from which all we see is but an effect which will eventually "run down" like a car which is out of gas. Thus they created and attached the *entropy* theory to the *Second Law of Thermodynamics* as if an integral part of it so that energy is constantly becoming "unavailable" as the universe runs down. And that would mean that the *Second Law of Thermodynamics* would violate the *First Law of Thermodynamics.***

**The missing ideas are how universal processes are continually self-replenished. Through an interplay of natural strong and weak**

forces acting with matter, perpetual motion is conserved. Strong forces overcome weaker forces and do more work in the same periods of time. Cosmic radiation travels through the ether, until it reacts with plasma, forming matter. When matter concentrates, forces acting with it concentrate. In contact with other, less concentrated matter, stronger cosmic forces acting within the space it occupies are unleashed, producing transmutations and eventually more cosmic radiation, which travels through space until absorbed and changed by plasma or other mass, producing larger masses or other radiation, and the process continues, ad infinitum. And all this reacting often with Tesla's Primary Solar Rays.

The rigid thinker must have a simplistic answer as to "where matter comes from", "where it is going", "what is the purpose of life", etc. Sometimes we must accept reality, and discard phony theories which attempt to give us a simplistic answer for what we either must admit that we just don't know, or should consider some questions too stupid to ask. The facts concerning the existence of universal matter, forces, vast distances, massive bodies and events, and perpetual motion are irreducible primaries or "first laws" about reality, which we must accept before we can progress with answering particular questions related to the more useful aspects of existence.

Since the orbital speed of the earth around the sun is greater than the speed of earth's rotation, our spot never moves "backward" relative to the ether, though it slows down and speeds up. The orbital speed of the Milky Way is greater than the orbital speed of the solar system or rotational speed of the earth, and our spot continues in a generally "wavy-wavy" path in which we accelerate and decelerate over a one-day longer cycle, and so forth. There are still larger and larger cycles of motion over time, as stated before. As can be seen from these cyclical movements, the universe is a system of interrelated systems which move generally in gear-like precision, rather than in a more randomized explosive pattern ala the Big Bang. Universal movement is in an orderly, cyclical pattern of circles (or ellipses), within circles, representing mechanical oscillations which are related to smaller and larger electromagnetic ones. The exceptions to this—the occasional cosmic explosion, such as that of a supernova—proves rather than disproves the rule. Even the occasional comet follows a cyclical pattern, while the debris from

**cosmic events eventually falls into orderly patterns. Though such explosions on our scale are gigantic, they are like minor ripples in a pond on a cosmic scale, and should not be allowed to warp our perspective. Neither are the events in the entire universe governed by the profuse scribbles written by someone on a fly-infested dung heap in the Middle East a few thousand years ago.**

**If last week a remote controlled telescope—such as the Hubbell —showed a movement which when compared to an earlier measurement of a few years ago indicated a movement away from a distant part of the universe, just remember that this is only an observation of movement taken from one perspective at one time of a much larger pattern of oscillations over a much longer period of time, in which the movement may later appear to be going the other way, from our point of view or that of the telescope, if we could make an observation millions of years into the future. Because of the circuitous patterns of movements which we observe in relation to earth at the present, and the larger structures in "our vicinity" of the universe, we have good reason to believe that all such movements are cyclical, and to extrapolate ourselves into the time when the appearance would be quite different. The theory that the earth and other celestial bodies are the result of the Big Bang are ludicrous. Just try to imagine the Big Bang explosion building the earth up from a small mass in space, to build thousands of feet of the detritus of living things, to increase the mass of the earth through photosynthesis, which has nothing to do with a Big Bang. What a coincidence! The universe has no "age", as it has always existed, and always will. The whole theory about how planets "were created" from the Big Bang is ridiculous.**

**Life forms would not only have to have existed at one time on Mars, but on all other planets and stars, at some time. In fact, the planets grew from smaller bodies as a result of life forms, which convert interstellar ether into solid mass, driven by cosmic radiation. There is no depth to which we can drill on earth, or otherwise produce direct evidence to the contrary, where we will not find matter which was in its past part of living things. This theory of course makes the planet "too old" for the Big Bangers, as it would take a much longer time for the earth to reach its present state, and would throw their "calendar"—called the "Hubbell Constant",**

**through which the high priests of the Big Bang have determined earth's alleged "birth date"—off.**

**For example, one of the components of chlorophyll is iron. This iron along with the carbon deposits and other elements result from photosynthesis. If strata of the earth, thousands of feet—even miles —thick, are composed of the decayed and metamorphosed remains of plant and animal life, the original biomass represents a conversion ultimately of ether, driven by radiation through photosynthesis, and into the many elements and minerals composing the strata, which can be easily traced to this process, such as iron, calcium, phosphorus, sulfur, nitrogen, carbon, hydrogen, oxygen, and helium. Such elements accumulate through the cosmically-driven forces of life forms. There is probably a constant of universal mathematical proportionality between life forms and inorganic ones. And we should realize that all of this mass of the earth's upper layers, down to the deepest level we can drill, represents mass which has been added to earth's mass by life processes in which ultimately cosmic radiation drives the accumulation of mass from the ether.**

**The ether, the medium which supposedly conducts sound waves in the ether through space, can apparently be converted into any form of matter. Otherwise, where would that mass come from? It is also apparently possible to precipitate any element from the ether, or to transmute other elements, using proper technology which emulates or speeds up this natural process. The ether contains the building blocks of what we call "solid matter", the proof of which enhances this possibility.**

**Tesla eventually realized the ether's use in his discovery of the "tremendous propelling force". His propulsion technology works and that is more important that splitting hairs...or atoms.**

**Only an ether which is intimately involved with momentum, could explain such a tremendous propulsion force such as that which I observed in 1953. This propulsive force was nothing more than a force which exists in nature, which in the hands of man becomes an article of will, once the physical processes capable of synthesizing and manipulating it become known.**

**Based on the assumption that I could see the ship for thirty miles, I estimate that from the rate and manner of acceleration, in conjunction with the probable mass of the ship which I observed, it**

was immediately apparent that the entire volume of the ship, if filled with conventional rocket or jet fuel, would have been consumed in the three seconds required for it to accelerate to ca. 36,000 mph and disappear at the point of infinity. This was only possible because of the existence of Tesla's technology which "electro-mechanically" canceled inertia and synthesized a new momentum almost instantaneously. This technology also created a friction-free pathway through the atmosphere and also prevented a shock wave or sonic boom, perhaps explainable by the fact that Tesla's "sound waves at the speed of light" are involved. Another explanation is that rapidly-varying negative electrostatic potentials apparently create a path of mutually repulsive, same-sign negative charges in the atmosphere before, along and behind the ship's path which do not clap rapidly back together behind the ship to create the sonic boom. I observed the atmospheric effects of the lingering charges scintillating and sizzling behind the ship, as they slough off.

What is inertia? Inertia is the momentum which a body "at rest" already possesses because it is in a state of uniform motion, but which to us appears to be at rest. Tesla's technology apparently uses the electrical attractive force, which produces $2 \times 10^{39}$ power times more work, using a smaller amount of "gravitational" or chemical "input energy" to drive an electrical generator to produce the electricity to trigger the change. If "energy" is the "ability to do work", and "work" is the "movement of a mass through a distance", and a stronger force will do more work over a given time period, a change in form of force from a weaker to a stronger force (which exists in the environment) will use "environmental energy" to do more work in a new way.

The process which makes Tesla propulsion possible, is the dynamism of the universe, which naturally exchanges weaker and stronger forces to conserve perpetual motion, with any lost momentum being re-supplied by the Primary Solar Rays or other natural cosmic radiation or forces. That seems the only way to explain such a fantastic demonstration of technology in defiance to so-called "accepted theory", but in accordance to Tesla's hidden ones which the government possesses in secrecy. Otherwise, why the secrecy?

## CHAPTER III: OBSERVATIONS

**With Tesla's theory and the background of older research and theories held temporarily in abeyance, let's look at what has been directly observed either by myself or by those closely associated with me.**

**What I began observing in 1946, and have continued ever since, are man-made electro-mechanical flying machines which can hover, turn on a dime at high speed without flying apart, and accelerate almost instantaneously to at least 36,000 mph, without readily audible sound or sonic boom. The ship I observed in 1953, in broad daylight, hovering at about 300 feet distance, about 250 feet above ground level, precessed at a high angle—about 45 degrees—and a low rate of about two cycles per second. Soon after I spotted it, it began to move away directly due west, quickly accelerating to about 200 mph, at which time it did two absolutely square turns—a right angle turn straight down and another right angle turn about a hundred feet lower, back into level flight—then accelerated to a point of infinity above the horizon about 30 miles away in three seconds. This maneuver to lower its elevation above ground by about a hundred feet appeared to be designed to make the ship less visible to observers on the ground, when the pilot discovered that he was hovering over a town. The high angle of precession appears in retrospect to have been a malfunction of the gyro-stabilization system, as it would have otherwise have made it very difficult for a viewer inside the ship to make any sense out of what he could see outside.**

**The ship was constructed as if it were two 50-foot diameter stainless steel woks joined together clamshell fashion. It had concentric "striations"—grooves in its surface which appeared to be about one inch wide all over its surface—which gave its two halves the appearance of giant "spun-metal" woks, as it precessed with the sun's rays playing on its oscillating bright metallic surface. The sun was about ten degrees above the horizon at the time.**

**As it precessed, the ship gave off a sparkling, "rainbow-colored" electrical corona, like a metallic surface which was being electrified with high voltage Tesla coils. On the shadow side of the bottom when it was periodically turned away from the sun during oscillations, there appeared a soft, 'cloudy', infra-red glow, which**

**indicated the possible presence of microwaves on the bottom, while the general "rainbow" corona from the top was "brushy" and apparently of high voltage negative electrostatics. As the ship began accelerating, the rainbow corona formed a sparkling, scintillating trail behind the ship, which also showed a "texture" across the trail which appeared to coincide with the frequency of the precessions which increased in rate as the ship accelerated, until a mere vibratory blur by the time the ship made its two 90-degree turns at about 200 mph, which is a rough estimate. The ion trail gradually faded out behind the ship, with little sparkles appearing in the trail as it did so. The ionizations produced by this ship—rapidly-varying electrostatic ones produced by the one-quarter wavelength tuning of Tesla's single terminal coil—apparently produced the sound waves which were projected along the direction of acceleration, and bringing in the tubes of force necessary for the propulsive force to act on the ship. All this theory was apparent from the effects graphically illustrated by the corona together with the actions of the ship.**

**The ship had flown due west, toward the northern border of Ft. Bliss, near El Paso, Texas. Ft. Bliss was the location of Headquarters, U. S. Army Ordinance, where Wernher von Braun was appointed research director in the summer of 1945. Von Braun's main base of operation had been moved to Redstone Arsenal, Alabama about 1951.**

**Consistent with the observed behavior, the electro-mechanical system appeared to instantly synthesize in an internal atomic way, inertia and momentum, things which normally create problems in acceleration and turning, but which the system overcame as shown by the evidence of the square, right-angle turns and extremely rapid acceleration. The system appeared to have very little to do with "gravity" in the textbook way, even though it did reveal that gravity itself is apparently due to the same kinds of electro-mechanics. These observations were taken at face value, as proof of certain facts concerning the nature of the ether, gravity, inertia, momentum, and electro-mechanical propulsion. It was also taken as proof of the fact that I saw a machine, that machines are man-made, and that Tesla had described his invention of a flying machine which had these properties.**

**I had observed whole squadrons of saucers flying at high altitude**

**in formation at night, beginning in 1946, and continuing in 1947 through 1953 at our next house, during which time I slept outdoors in our back yard. The squadrons arrived over our town from the direction of New Mexico, the border of which was about ten miles north. The ships departed back to the north after their night-time maneuvers. Ergo, those particular ships were from a New Mexico base, which I took to be Holloman AFB at Alamogordo or White Sands Missile Range.**

**The night squadron saucers turned in angles of 30 degrees or multiples (30, 60, 90, 120, 180, etc.) which meant those particular saucers had the same navigation and control system that my 1943-vintage Peiltochterkompass was used on.**

**My observations, together with all the written information on the Peiltochterkompass, and an analysis of the device itself, revealed quite a lot about the German "Kurskreisel Tochterkompass-p2" and the propulsion system:**

**1. The ships were electro-mechanically propelled;**

**2. the propulsion system created an electric field around the saucers which necessitated the development of a celestial guidance system and compass, since a magnetic compass would be useless due to the "Faraday cage" effect;**

**3. the ship was a circular, discus-shaped craft, which turned in 12 peripheral directions (thirty degrees each) of the compass, plus "up" and "down";**

**4. the ships used a celestially oriented master navigational gyro ("Kurskreisel"), calibrated prior to take-off to the Pole Star, Arcturus, as the frame of reference with which to navigate the ship using an electromagnetic slave compass ("Tochterkompass"), which was also interconnected to stepping switches which actuated the ship's bi-polar propulsion electrodes as the pilot wished;**

**5. the system canceled the normal effects of gravity, inertia, and momentum, and instantly synthesized momentum in the chosen direction, without disturbing the operation of the master gyro, which continued to maintain its heading during flights, apparently because of its 'internal' inertia/momentum within the ship's outer shell;**

**6. the instant turns showed an incompressible ether which is integrally locked with mass in the phenomena of inertia, momentum, and gravity.**

**The celestial orientation of the *Kurskreisel* to the Pole Star meant that no matter where on earth the ship flew, true north would be correctly indicated. If a line is drawn from the Pole Star down through the center of the earth, this is the axis around which the earth revolves. This master gyro had electrical contacts around it's gimbal so that as the ship was turned in the 30-degree increments, the contacts would relay that information to the slave compass so that the small motor would come on and rotate the geared flight direction indicator ring until it showed the correct direction of flight in relation to true north.**

**Since the 1953 daylight sighting, I have seen many other electro-mechanical craft, but one of the best sightings I made was on a moonlit night while soaking in a hot tub.**

## Rub-a-Dub-Dub, Seven Folks in a Hot Tub

**On a Wednesday evening, January 24, 1996, at approximately 7:00 PM, I sat in a hot tub with seven other people, at the Ten Thousand Waves Japanese Bath House, located on the western slopes of the Sangre de Cristo Mountains, in the Hyde Park vicinity above Santa Fe, New Mexico. As I steamed in the tub, conversation between a couple from Denver and myself drifted from the dangers of fluoride to "What do you do?", to which I answered, "I write some". Asked what I had written, I answered "A book called *Space Aliens From the Pentagon*".**

**A beautiful woman on my right exclaimed "I have your book. I'm a friend of Alex and Rex. I love your book so much, and so do they." I thanked her and we exchanged introductions, and then I said "I see saucers up here almost every time I'm here after dark...and as a matter of fact, there's one right there" (pointing out a flashing aircraft flying slowly easterly about 15 degrees to the lower left of the crescent moon, at a distance of about 10,000 feet from us).**

**A heckler across the tub said "Aww, that's a plane", to which I responded, "There's another one...and if they're planes, why have they parked? (The ships had become stationary in the air.) The flashing lights mask an underlying glow. The eye recoils from a flash, and never has time to recover before another flash. When they move**

**faster, they glow brighter, which is why they are going so slowly."**

**No sooner had I spoken than the stationary ship on the left quit flashing, and turned into a much larger dark, glowing green shape like a football field surrounded by a track. Around the "track" was a dark strip, with a purple corona outside that, produced by the Tesla coils. I could see the curvature of its bottom as it moved, flying in a large circle, as I followed it with my outstretched arm and index finger, and my fellow hot-tubbers oggled in jaw-dropped amazement. The following is a black-and-white photo of a painting I did of the saucer, as seen from the bottom. The ship is moving upward in the illustration, in the direction in which the electrostatic leaders are moving:**

**It returned to its starting point, stopped instantly and began**

**flashing like a small plane again. This took approximately ten seconds. Since the ship appeared about three-quarters of an inch long at arm's length (about 2 feet), at the 10,000-foot distance, it was around 400 feet long and 250 feet wide. Since the diameter of the circle it flew in was about 30 degrees, its speed was about 2,000 mph (the speed didn't surprise me since I had clocked the one I saw in 1953 over our back yard at a minimum of 36,000 mph when it departed; I had also clocked one on a radar scope at Keesler A.F.B., Mississippi in 1956 doing 4,000 mph flying over Lake Ponchartrain).**

**I almost missed this ship because I didn't have my glasses, and because it appeared to have been "stealth-ized"—designed to be concealed by the dark night sky, by controlling its emission color to be dark green, with the purple corona also being dark, so as to more closely match the light value of and blend into the night sky. The main 'give-away' was the color differences.**

**Others in the tub—most of whom didn't need glasses—saw it more sharply than I, yet wouldn't have seen it at all had I not pointed it out to them. I probably would not have seen it myself had I not been observing as a habit, the flashing ships beforehand, with the moon so close above, creating some back-lighting which helped to define the larger, darker silhouette when it arrived. Moreover, the real reason I saw it, was my own preconscious mental self-conditioning.**

**I was amazed at its large size, since, when flashing, it appeared like a small plane with beacon. When I finally looked back to the other side of the tub, the heckler was gone, the on-cue saucer demo being apparently too much for him. If the saucer crew had been listening to my words, their performance couldn't have been more perfectly timed. Bravo! Do they also have some computer software that reads lips? It is my theory that the human saucer-jockies have chosen that particular location to do some "testing", because of the titillating availability of the numerous private hot tubs open to the night sky, occupied by private nude couples of all sexual persuasions, easily video taped by them. I can see them trading the tapes among themselves, another good reason to "...keep the secret" to avoid prosecution for invasion of privacy and violation of voyeurism laws.**

**It was the first time I had seen the green-glowing type reported years ago by astronomer Clyde Tombaugh, then recognized as the**

**discoverer of the planet Pluto. It was the second most graphic demonstration of the saucer technology I have witnessed. If I had not spotted it, it is probable that none of us would have seen anything extraordinary, excluding the beautiful lady next to me who liked my book, and was a strong competition to the saucers. Incidentally, the female breast shape may be "natures most nearly-perfect shape for a flying saucer design." The "green-glowing" type was apparently the result of research designed to make the ships less visible in the night sky, since the early models—such as the "foo-fighters" ("Fliegende Schildkrote" [Flying Turtle] and "Kugelblitz" [Ball Lightening]) glowed brightly. The oblong design (as distinguished from the circular or 'peripheral' saucers) comports with what I call the "linear type". According to a friend who is an expert spectrographer, the green corona indicated atomic absorption, which I presume to be 'stealth' technology used to suppress the white light spectrum. He seemed to think sodium based coatings, while I think maybe chromium, because it appeared to be a veridian ("chrome") green. The soft, 'cloudy' green also indicated positive corona, as distinguished from the 'hairy' negative (purple) corona around the outer periphery, and usually seen on Tesla coils.**

## Comparison to Other Sightings

**Though remarkable, the hot tub sighting was not a clear, "daylight", "close-range" sighting, like the first good sighting which I had personally made on a bright, late summer afternoon in West Texas, in 1953, within 250-300 feet, right over our back yard, in Kermit, Texas, as our family and friends were having an ice cream party. I had been the last in my family and neighborhood to see a saucer in a close-up daylight sighting, since three had appeared on two consecutive days, June 1st and 2nd, 1950, while I was unfortunately in the Davis Mountains about 120 miles to the west. My father in his lifetime had seen more than I had, and was once held in his car for 15-30 minutes, on the night of Nov. 2, 1957, just south of Levelland, Texas, by one of the 200-foot-long ships, which had scared the livin' daylights out of him. The thing hovered within ten feet of his car top, killed his engine, and burned out all his lights,**

yet failed to burn out a single fuse in any of the car's circuits. The saucer's electric field was apparently so strong that it arrested current flow, and was strong enough to cancel the magnetic fields in the generator and circuits. The white-hot tungsten filaments of the bulbs were the weakest points in the circuits, and the saucer's pulse currents apparently subjected them to magnetic and electric shear forces which broke them apart.

My father had to drive back to Levelland in the dark, spend the night in a motel, then drive home to Odessa, Texas at dawn's first light. He arrived home about 7:30 AM, had a quick breakfast, and left for Sewell Ford to have his lights fixed, saying nothing about the saucer. I turned on the TV for the morning news as he left, and realized that he had been one of the several drivers held by the ship (or ships) that night. I went to Sewell Ford and questioned him about it as the mechanic fixed the lights, and he asked me how I knew about the ship, and I told him it was on the news. He said he didn't want to be called "crazy", so had remained silent. This was what I call "anti-paranoia". It is sometimes reasonable to fear ridicule created by false government propaganda.

Later, in 1963, I had the opportunity to discuss the matter more fully with the sheriff of Hockley County, while working in Brownfield, Texas. The ship seen by my father matched in shape, appearance, and performance, the type observed by deputy Lonnie Zamora, near Socorro, New Mexico. Zamora had at first described the crew of the ship he saw as typical G.I.s in gray-green Air Force flight suits, but later altered his story, saying they were "...about the size of children". This was after the security spooks had 'leaned' on him.

In 1993 to 1998, ships could be seen almost any night in the area where I live in New Mexico, and were very profuse in an even more obvious way around the Cheyenne Mountain area to the south of Colorado Springs, Colorado, where the underground NORAD complex is located. The prevalence of the ships near my home diminished rapidly after my books became more widely read, as this caused a lot of people to begin observing them.

There has been another separate problem which I previously thought to be unrelated, the fact that I have been working on a book about ancient petroglyphs and have been blocked out of the San

**Cristóbal Ranch which I had free access to for over sixteen years before it fell into the hands of Ronald Reagan's blind trust. The trustee was John Henry Singleton, who eventually bought the over 81,000-acre ranch, which has some of the most important petroglyphs I know of for my research. The gluttonous ranch holdings for "Singleton Ranches" (there are several in New Mexico and some in California) had apparently become the receptacle for the large corporate income derived from military contracts by the two military-industrial complex corporations which Singleton was co-creator of, Teledyne and Raytheon. I had always believed that the reason I was blocked out of the ranch after the change in ownership was because of the archeologists, who hate my competing discoveries, and don't want them published. I now know that the reason is additionally and especially due to my UFO research, made apparent after I gave them a copy of *Space Aliens From the Pentagon* in 1994, after which I was immediately and threateningly blocked out of the ranch. I am humiliated by the shoddy and rude treatment and I have never encountered such inhospitable behavior by land owners in the Southwest---who are usually dignified and hospitable---and maybe that is because these corporate, military-industrial complex-type people from Los Angeles just don't fit in here.**

**In January, 2010, as I was watching TV in the dark at 2:00 A.M. because I couldn't sleep, I spotted a large discoid ship passing by through the large window above my TV on a downward path toward the ranch. Two minutes later I saw a second ship identical to the first one on the very same glide path toward the ranch. In the following nights when I traveled down the highway going home from Santa Fe I observed ships coming out of and going to the ranch. There is a saucer base over there. I should have figured this out before since Ratheon and Teledyne, and some of their many subsidiaries, manufacture components for UFOs. Their security there is way too high to be explained by mere "archaeological digs". They don't want me over there because they fear I'd discover the base. And now I have.**

**With the several direct observations or reports from those in my family or from close friends kept firmly in mind, let's see what more can be found in ether theory history to reflect on them.**

## The Ether and "Ponderable Matter"

In 1879, James Clerk Maxwell (1831-1879) said that the velocity of the solar system relative to the ether could be determined by observing the retardation of the eclipses of Jupiter's satellites. The purportedly observed behavior—in which the light of a star can be viewed with a steady image as the earth moves through space—indicates a stationary ether relative to a moving earth, solar system, and other bodies. But according to Tesla, this idea was defective, since you could not make such an observation on earth because the ether within earth's electric field is carried along with the earth. You might be able to make such an observation in outer space.

The ether is ultra-fine and presumably composed of positive and negative electric matter, which pervades all so-called "free space" as well as space "occupied" by "mass"—which the old ether physicists called "ponderable matter"—even though mostly "space". Since the ether is stationary near the earth, it is 'ponderable matter', at least where the ether is rigidified by earth's rapidly-varying electrostatic potentials and electric field.

The ether is transparent and does not refract or reflect light because of its "inconceivable tenuity" (Tesla said), although it is said to behave as a solid to light and heat. To "access" the ether in a propulsive way, rapidly-varying high voltage electrostatic potentials are required, in accord with Tesla's sound waves in the ether theory, first confirmed by the tests conducted by Tesla in 1891 (later detailed in his 1891 lecture before the A.I.E.E., Columbia College, N.Y.). Tesla's tests confirmed a reaction which was more than the feeble ionic reaction of light gases or so-called "electro-gravitics" such as those that the misinforming 1950s "patents" of T. Townsend Brown, later used in attempt to induce the gullible public to believe was the correct technology. After his tests, Tesla stated that the ether became a "solid state" medium which could be accessed by subjecting it to electrostatic impulses of "sufficiently high voltage and frequency".

The ether also becomes more apparent when there is a sudden change in the direction of motion, rate of acceleration, or velocity of a body. The relative movement of a charged body, equivalent to current, apparently creates a new degree of electrical displacement of the ether through which it moves, and influences the resistance to

**changes in velocity through the affected ether by the protons, neutrons and electrons composing the body itself. The increased work required to move such a charged sphere—the increased "resistance to change" called "inertia"—reflected by the increase in its tendency to remain in its current state of motion (actually its current "momentum"), demonstrates that the changes have created the resistance to movement of the sphere through the space through which it moves. Therefore, once in relative motion at a constant velocity—though more work was initially required due to the increased charges to give it that increased relative motion—the sphere's momentum should be increased consistent to an increase in its "virtual mass". This should make it clear that both inertia and momentum are caused by the same thing, which is a persistence in the reaction patterns in the atomic and electrostatic structure of the mass to maintain its present state of motion due to its properties as they are affected by the ether. Just as changing magnetic fields or electric fields are required for inductance, the creation of such changing electrostatic fields are the basis for producing the sound waves in the ether for Tesla's electro-mechanical momentum.**

**All bodies are in motion, and within all bodies, protons, neutrons and electrons are always in motion, because "elasticity" and vibratory motion occurs in "ponderable matter" even when a body is at relative "rest". The ether apparently sets up certain patterns of motion in ponderable matter passing through it, relative to that motion, and changes in those patterns brought about by electrostatic inductance can impart new motion to that matter, or inhibit its motion as shown by the charged sphere. It is clear that something in space affects motion, otherwise it wouldn't be harder to move a charged sphere through the space so engendered.**

**A body at relative rest on the earth sits within the ether rigidified by earth's rapidly-varying electrostatically-induced mechanical waves, and there is a down-force on that body called "gravity" created by the tubes of force being pulled down. As a body on earth moves parallel to earth's surface through the ether within its electric field, rigidified by earth's rapidly-varying electrostatic potentials (creating sound waves in the ether), it accumulates some of these ether "tubes of force" commensurate to its motion, which gives it added horizontal momentum.**

**It is possible that the tubes of force, induced to flow into an earth-bound body, can cause a downward thrust on the body's atoms, away from the pre-existing "equilibrium position" (which the atoms would occupy in the absence of earth's electric field) which we call "gravity". The equilibrium position for the electrical particles of ponderable matter varies according to mass density and displacement of the ether. A change in this electric displacement may result in a new equilibrium position. The body always moves toward equilibrium. The time-relative aspect of the rate of acceleration of a falling body is apparently due to its relative accumulated and dissolved tubes of force as it falls, regardless of its mass density.**

**In summary, inside an earth-bound body at relative rest there is a difference in ether displacement effects from that which exists outside the body in free space. Since earth's electric field is vertically graduated through the intermediate space between the negative at earth's surface and the positive at the ionosphere, the electric properties of mass are subjected to graduated strains. The effect of electrical displacement and gravity vary accordingly, somewhat like Newton's "inverse square" law. It is here that a difference between the "force of gravity" and the effects of inertia and momentum become apparent, since outside the earth's electric field there is apparently no gravity.**

**Traditionally, inertia has always been considered (according to Newton's First Law of Motion) the tendency of a body "at rest" (actually in "constant velocity" with earth) to continue its state of rest (or constant velocity), unless acted upon by an external force. Since all bodies at relative rest are already moving at the same "constant velocity" of the earth however, only changes in movement relative to earth are at issue, and any such changes affect the patterns of flow within mass of the tubes of force, as the body moves through the rigidified ether, at any given time. Since only a body which moves relative to the earth has been traditionally said to have "momentum" ("the product of a body's mass and velocity"), this rule ignores the fact that all bodies "at rest" relative to earth already have "momentum" therefore "inertia". And Tesla turned these theories upside down.**

**Momentum and inertia are apparently due to the exact same "tendency", which relates ultimately to a frame of reference to the**

**universal ether, but for convenience—since we don't have an absolute "fix" on the ether—we use the moving earth as a frame of reference from which to measure a body's rate of change relative to the ether, as the earth maintains it relatively constant velocity. There is a special case on earth, because the ether within earth's electric field has been "conditioned"—that is, it has been subjected to electric strain. This comports with Tesla's statement concerning the effects of "rapidly varying electrostatic forces" emanating from the earth.**

**The earth has momentum within the sun's frame of reference, just as a moving car has momentum within the earth's frame of reference. Some physicists have mistakenly applied a "gravitational field of influence" idea, thinking this somehow isolates a body from references outside this field. I believe that this is only an "apparent effect" and that earth's "gravitational field" is a product of its moving electric field (a "current") as it influences the ether within it, and that the so-called "gravity field" extends no further than earth's electric field.**

**Momentum is due to the accumulation inside a body of the dissolving tubes of force as the body passes through the ether, or in the case of UFO propulsion, the electro-mechanical inducement of the inward flow and dissolution of the tubes of force imparting mechanical momentum to the ship. The rotation of these tubes of force is the product of the magnetic rotation of the plane of polarization first identified by Michael Faraday. Any variance from the earth's uniform motion results in a resistance to change in the orientation of the tubes of force, since the rotating tubes of force spin around their irrotational ether cores over a given time and a change in motion forces them to rotate around a greater or lesser quantity of ether depending on whether the change is an acceleration or deceleration. The tubes possess electro-mechanical force which is at least $10^{40}$ power times stronger than the gravitational interaction (or 2 X $10^{39}$ power times stronger than the so-called gravitational attractive force), which explains why these tiny tubes can have such a great influence over a mass in motion. This force is locked into the discrete flow pattern between the ether and the tubes, which are coupled to the 'ponderable matter' structure which is the recipient of the momentum.**

**Though the flow patterns are always gradually changing because**

of the continual reorientation of the earth as it moves in its angular momentum around the sun and in its motion with and within the solar system, these changes are sensually imperceptible because they are so gradual over time, with effects which are virtually constant. Though the changes are too slow to sense, the action of the tides are related to them as well as the electrical displacements caused by the positions of the moon relative to earth.

The natural gravity force is always normally toward the earth, which is negatively charged, within earth's electric/gravity field, due to the polarity of that field, since the negative component of the ether in free space is electrically displaced upward by the repulsion of like charges. Meanwhile the inertial characteristics of positive mass is forced and attracted downward, and is more resistant to the changes in displacement.

The movement of ether particles is related to the gravity force. Due to the greater velocity of the electromagnetic interaction at C however, this down-force is irrespective to the direction of earth's movement since the interaction between the electrical components of matter and the ether due to the conditioning by the earth's electric field, is vertically polarized. This creates the same proximate effect on mass over time, irrespective as to earth's orientation relative to the ether external to the earth's electric field.

## Orbiting Planets and Satellites

The concept that a satellite in orbit around earth is "held" in orbit by "earth's field of gravity" seems false. The logical implication of my analysis is that the orbital path taken by a body is because it "sees" that path as one of least resistance, as if it were a straight and level path. Since for a given radius above earth the ether is displaced to a certain degree, an orbiting body would experience an accelerating force which would be met by inertia if it were to move upward, and a decelerating force which would be met by momentum if it were to move downward. Its path in force-free constant velocity remains at a constant radius or level of ether displacement, as the path of least resistance.

Since the determinate factor which limits a body's motion is the reactance of its internal electrical content with the ether at its

**particular degree of displacement by earth's electric field at a given altitude, an orbiting body "sees" its orbital path along a path of like displacement. This is contrasted to how Einstein said that space and time are "wrapped" around a large spinning body as it moves through space because, without an ether, there would be nothing in "empty space" to wrap, and time cannot be wrapped because it doesn't exist as an entity, being our human concept invented to measure the rate at which events occur.**

**One of the revealing facts obviated by films of astronauts floating in the so-called "gravity-free" space of an orbiting space station, is the contradictory assertion that the space station is "held" in orbit by earth's gravity. If earth's gravity exerted sufficient force to hold the station in orbit, the internal environment would not be "gravity free", but would pull the astronauts down to the floor, instead of allowing them to float all around inside, as we've seen so many times on TV. Yet, there is sufficient elasticity within the internal masses of the astronauts' bodies, between electrical content and the ether, to allow movement in that manner.**

**If one calculates the phenomenon on a scale model of the earth one foot in diameter, the orbits of the astronauts would be only a tiny fraction of an inch above the surface. For there to be "no gravity" at that "altitude" shows the ridiculousness of the loss of gravity and refutes Newton's inverse square law. But after all, Boscovich was the father of Relativism and Newton's corpuscular theory of light was the basis for the eventual elimination of ether theory from textbook science.**

## CHAPTER IV: HISTORICAL DEVELOPMENT OF THE THEORY

### Rotatory Motion and the "Screw Effect"

**Wm. Thomson (Lord Kelvin) asserted that magnetism possesses a "rotatory" character related to heat or the thermal motions of a body [17]. Nikola Tesla made many references to Wm. Thomson, pointing to his work as a prelude to his own discoveries and**

17[17] Wm. Thomson, Proc. R.S. viii (1856), p. 150.

**applications which especially intensified in 1892. A review of the work of the world's major minds leading up to Tesla's breakthrough is necessary to show just what Tesla discovered and what it meant in respect to ether physics and physics in general. This is a task made more difficult by the concealment of Tesla's papers by the U. S. government. And incredibly, the suppression of Tesla began before the 20th century by the absence of references to Tesla or his work by the "Good Old Boys' Club" in Europe.**

**Around 1870, Thomson had conducted experiments which seemed to indicate that "gravitational action" could be induced by spheroidal bodies oscillated by electrical currents or mechanical pulses[18]. The surface pulsations could cause attractions or repulsions in respect to other bodies, as verified by Thomson. Tesla was aware of Thomson's work during his student days in Graz, Austria, beginning in 1875 when he was nineteen. Thomson's work undoubtedly served as a spark of inspiration for Tesla in his early conception of an "ideal flying machine" which would be propelled by electricity acting upon the ether. This explains Tesla's continual references to Thomson, such as demonstrating during his 1892 London lecture a 'luminous wire' sign powered by a Tesla coil, which said "WILLIAM THOMSON".**

**At first, Thomson found that ponderomotive forces act between two solid bodies immersed in an incompressible fluid, when one of the bodies is immobilized and made to oscillate with a force which acts along a line between its center and that of a much larger sphere which is free. The free sphere was attracted to the smaller (immobilized) sphere, if its density was greater than the fluid, while a sphere of less density than the fluid was repelled or attracted, according to the ratio of its distance to the vibrator in relation to a certain quantity.[19]**

**Thomson's experiments were analogical ones, for which he had evoked praise from his contemporaries even when he was still a teenager, although his refusal to believe anyone's assertions unless he could build an analogical model to prove them often led to the consternation of his contemporaries, such as Maxwell, who relied**

---

18[18]. F. Guthrie, Phil. Mag. xli (1871), p. 405.

19[19]. Wm. Thomson, Phil. Mag. xli (1871), p. 405; Letter, Thomson to Guthrie, p. 427.

**often on mathematical equations. The sphere experiments were designed to use mechanical and electrical wave methods to construct a model to probe the gravitational, inertial and momentum reactions of solid bodies in the ether.**

**The Faraday Effect—the rotation of the plane of polarization of radiation in a dielectric medium (such as the atmosphere and certain solid materials) in a magnetic field—stated that the angle of rotation of radiation is proportional to the magnetic field strength and the length of the path in the medium in the field. These early experimenters also knew there was a connection between the rotatory motion and momentum, and sought to find it.**

**The rotatory (versus the linear) character of magnetic phenomena was strengthened by Thomson's experimentally verified conclusions on the magnetic rotation of light. This rotatory character not only influenced Tesla's discovery of the rotating magnetic field, but is also fundamental to inertia and momentum, as I will later explain, since movement of a charged body constitutes a current which creates a magnetic field which creates the rotatory motion which hypothetically may "bore" through momentum-carrying ether particles to create electromechanical momentum.**

**Thomson's system was later investigated by C. A. Bjerknes between 1877 and 1910. Bjerknes showed that when two spheres immersed in an incompressible fluid were pulsated, they exerted a mutual attraction which obeyed Newton's inverse square law if the pulsations were in phase, while if the phases differed by a half wave, the spheres repelled. At one quarter wave difference, there was no action. Where pulses were non-instantaneous at distances greater that a quarter wavelength, attractions and repulsions were reversed.[20]**

**These conclusions were hypothetically related to possible propulsion. The publishing of these researches and experiments in the physical journals of Europe were available to Nikola Tesla during his student days at the Polytechnic Institute in Graz, Austria, and at the University of Prague, in Czechoslovakia. Tesla could read and understand all these pertinent journals in their original languages.**

**Around 1878, George Francis FitzGerald (1851-1901) [21]**

---

20[20] C. A. Bjerknes, Repertorium d. Mathematik I (Leipzig, 1877) p. 268; Proc. Camb. Phil. Soc. iii (1879), p. 276; iv (1880), p. 29.

21[21] George Francis FitzGerald, Phil. Trans. clxxi (1880), p. 691; FitzGerald's Scientific

**compared magnetic force and velocity in a quasi-elastic solid, based on a model devised earlier by James MacCullagh (1809-47)[22], whose model was the only one which could propagate waves with the properties of light. This was obviously analogous to the electromagnetic theory of light—as shown by MacCullagh's ether equation of motion and ether theory which made it feasible to extend ether concepts to represent optical phenomena, along with magnetic and electrical interaction. According to Tesla the electromagnetic theory of light was erroneous because the ether is not a solid capable of transmitting transverse waves. To Tesla, light is also sound waves in the ether and is mechanical and not electrical or magnetic.**

**An Electrostatic Charge Carried Around**

**In 1879, Edwin H. Hall[23], a student in Baltimore, repeated an experiment suggested by H. A. Rowland, his professor, whose original experiment with a gold-foil-covered ebonite disk in a magnetic field showed that electric charges on a disk were carried around with it as it was rotated.[24] In Hall's experiment, a gold leaf strip in which a current was flowing, was placed into a magnetic gap. This produced an electromotive force at right angles to the magnetic field and the current, which was proportional to the product of the two. Called the "Hall Effect", it was already inherent in the three previous effects discovered much earlier by Faraday.**

**Faraday had discovered induction, by forcing a conductor through a magnetic field, cutting the lines of magnetic force and producing a current in the conductor, and also produced a magnetic field in an unmagnetized iron core, by forcing a current-carrying conductor through a gap between the poles of the core. Faraday's third effect was the generation of a current. Though Hall's effect was inherent in the fact that it was the reverse of the force required by Faraday to push the conductor through a magnetic field, Hall's work completed the triad of effects, by experimental verification. This electromagnetic principal applies to solids not to the ether in free**

---

Writings, p. 45.

22[22]. James MacCullagh, Brit. Assoc. Rep., (1835).

23[23]. Edwin H. Hall, Amer. Jour. Math. ii (1879), p. 287.

24[24]. H. A. Rowland, Ann. d. Phys, clviii (1876), p. 487.

**space. This is the basis for MHD (Magneto-Hydrodynamic) generators, and is integral to fluid propulsion.**

**Since the galvanometer needle in Hall's experiment was deflected only when the magnetic field arose or collapsed, the physical thrust created was a vector product which had already been expressly suggested by Maxwell[25] almost 15 years earlier (derived from Maxwell's analysis relative to Faraday's work of ca. 1845), though Maxwell failed to follow up with experiment (because he died), but the equations are still used.**

**Though it was said by Whittaker that the Hall Effect, like the magnetic rotation of light, occurs only in ponderable bodies and not in the "free ether", this statement is patently false since the effect actually depends on the conductivity of a medium. The fact that the effect occurs in 'ponderable bodies' and 'conductive media' however, seems important for electro-mechanical propulsion, since it shows the reaction between such bodies and media and the underlying "ether framework" which is accessed in the process.**

**Since the "natural media" (the ether and the atmosphere) so often referred to by Tesla in his patents become conductive under the influence of an electrostatic force of sufficiently high voltage and frequency, the effects of electrostatics on the free ether, dependent upon proper conditions, can impart momentum to a ponderable body, so as to move it though the free ether. A startling proof that Tesla's sound waves work in the free ether, was Tesla's electro-mechanical plate suspensions in his laboratory in 1898.[26] This was also verified by Tesla's transmission of mechanical energy through space by longitudinal sound waves which travel at C, as explained in his 1892 Lecture before the Institute of Electrical Engineers, London. This also showed Tesla's priority with radio discovery.**

**Since an electric field 'displaces' the ether—which is the basis for MHD pumping (especially when pulsed)—the effect actually showed an operable "electromotive force" ("emf"), or "electro-propulsive**

---

25[25.] G. C. Maxwell, Treatise, (1862).

26[26.] F.B.I. Papers, (1943), disclosed through the Freedom Of Information, in the statement of Marguerite Merrington. The FBI disclosure documents also show that the contents of Tesla's safe, two truckloads of papers and apparati, 75 packing crates and trunks and another 80 large storage trunks were all removed and stored in the warehouse of the Custodian of Alien Properties in New York in 1943.

**force", between ponderable bodies and the ether, by means of electro-mechanical action. But the high voltage and high frequency are required to 'access' the ether. Tesla had succeeded in transmitting mechanical energy by means of high voltage, high frequency electrostatically-induced sound waves —"radio waves"—'accessing' the ether. Tesla's work at that point had already verified experimentally everything that he said and refuted what Maxwell had "mathematically" analyzed as being the supposed "electromagnetic" nature of light ---Faraday's theory---which is false because the ether is not a solid but a gas.**

**Though it was implied, the literature available to me failed to explicitly state the idea that inertia and momentum are the products of an electrostatic force acting on the ether which acts within bodies and all space. Neither was it specified that a pulsating sphere or other ponderable body can be propelled through the ether, without the presence of another sphere or ponderable body to pull against— except in the statements of Nikola Tesla and his "Flivver" electro-mechanical "ideal flying machine".**

**In 1884, the year Tesla invented the rotating magnetic field, J. J. Thomson attempted to determine the field produced by a moving electrified sphere, and the mathematical development of the theory accelerated. It was naturally easier to solve such problems from the known behaviors of simple geometric forms—planes, spheres, and cylinders.**[27]

**The possibility that the ether was composed of positive charges carrying their own 'sub-quantum' negative charges which were elastic and could be displaced had apparently evaded the thinking of Thomson, who had discovered the electron, the charge of which was considered indivisible. Although he had assumed that displacement currents must occur in the ether, he had earlier thought this was due to the magnetic effects of moving charges, though he failed to show how the displacement currents occurred, or what their specific effects were in terms of inertia and momentum, until after Tesla's 1892 lecture in London, which he attended.**

**There was already a sort of battle brewing between the proponents of classical electrodynamics and the proponents led by**

---

27[27.] J. J. Thomson, Proc. Lond. Math. Soc. xv (1884), p. 197.

**Maxwell of an electromagnetic theory of light. To the former, conductivity occurred in metal wires, etc., while with Maxwell, it occurred in the surrounding dielectrics and ether-filled space, with the conductors serving only to "guide" the action. Tesla departed from the Faraday/Maxwell camp, particularly on some major points such as a solid versus a gaseous ether. Maxwell's work seems valid in terms of electromagnetic effects in solids. FitzGerald attempted to unify the two views by arguing that Maxwell's unification was valid because radiation could be generated by purely electrical means. It is interesting to see how these researches contrasted with the development of the electro-mechanical propulsion of Tesla.**

**Along this line, Wm. Thomson (1884) first considered a charged sphere moving uniformly in a straight line. He assumed that the electric charges were uniformly distributed, with an electric field the same in all directions, no matter what position the sphere was in, the same as if it were at rest. This assumption proved true so long as the velocity of the sphere and the velocity of light were neglected.**

**In 1889, Wm. Thomson stated[28], "Rotational vortex-cores must be discarded; and we must have nothing but irrotational revolution and vacuous cores." By this, Thomson meant that the vacuous "ether" inside rotating tubes of electromagnetism did not rotate. Perhaps meaning that if the cores rotated along with the rotating tubes of electromagnetic force, it would neutralize the electro-mechanical action by which the tubes of force move.**

**FitzGerald found a purported error in Thomson's work, saying that the required "circuital condition" was not satisfied unless the moving charges on the sphere were considered as current, combined with the displacement and convection currents due to the motion. In correcting Thomson's error, FitzGerald went overboard in concluding that the magnetic force due to the displacement currents of the moving sphere had no resultant effect.**

**In 1888, Oliver Heaviside showed that the electrostatic units "vanished" to the inside of a moving sphere. This was the opposite to Faraday's experiment in which electrostatic charges placed inside a stationary, closed vessel "appeared" on the outside. Apparently, movement of the sphere—which increased its momentum—appeared**

---

28[28]. Wm. Thomson Proc. Roy. Irish Acad. i (30 Nov. 1889), p. 340.

**to force the charges back inside.**

**Heaviside's conception of the "spherical" symmetry of charges during movement was disproved by G. C. F. Searle in 1896[29]. Searle found that a moving "point charge" system is not a sphere, but an oblate spheroid, with a polar axis along its direction of motion. What Whittaker failed to point out was the importance of this finding, a connection between inertia, momentum, current, surface charges "vanishing" and "reappearing", and an electrical polarity along the direction of momentum, as well as an electro-mechanical link between the movement and the ether, since the displacement of the electric lines and polarity correspond to the movement, consistent to the thinking that the tubes of force deliver momentum to a body forced to move through the ether.**

**During this time, Nikola Tesla had not tarried. He had already shown that the "circuitous condition" could be met in a totally new way. In his lecture before the A.I.E.E. at Columbia College, N. Y., may 20, 1891, he demonstrated his years-old technology, and stated that he connected "one terminal" to a lamp and the other to "an insulated body of the required size. In all cases the insulated body serves to give off the energy into the surrounding space, and is equivalent to a return wire." In this lecture, Tesla also spoke of his electro-mechanical momentum theory which was similar to J. J. Thomson's discovery of 1893[30], which was flawed.**

**In the same year as Searle's finding (1896), W. B. Morton[31] similarly showed that the surface density of a charged body is unaltered by motion, but the lines of force no longer leave the surface perpendicularly (like the ship I saw while hot-tubbing). He also found that the energy of the surrounding field is greater when in motion than when at rest. Since greater work is required to move a charged sphere, than an uncharged one, with no increase in the actual mass of the sphere, a connection between moving charges and an ether was verified. This was considered true because the charges increase the "virtual mass" of the sphere, and the self-induction of convection currents is formed when the charges are set in motion by**

---

29[29.] G. C. F. Searle, Phil. Trans. clxxxvii (1896), p. 675.

30[30.] J. J. Thomson, Recent Researches in Elect. and Mag.., (1893), p. 13 (Supra.).

31[31.] W. B. Morton, Phil. Mag. xli (1896), p. 488.

**movement of the sphere, but neither of these explanations seemed to explicitly note that a force between a moving charged mass and the space through which it moves must have an ether framework to push or pull against, or that a current is caused to flow between matter and the ether, (as in Tesla's "circuital condition") due to the movement.**

**J. Larmor[32] suggested that the inertia of ponderable matter may be ultimately proven to be of this nature, since atoms were constituted of systems of electrons. If a greater "virtual mass" effect (W. B. Morton, supra) is created electrically, which increases or decreases the ease of movement of a body through the "free ether", and increases the total energy of the moving system, then a link between ponderable bodies and the ether framework was proven. If the increased charges opposed a body's movement, then the system could be reversed to produce a thrust on the body. This suggested that a system of movement of charges into or through a sphere would propel it through ether-filled space.**

**There were similar implications in the works of Faraday, Maxwell, Wm. Thomson, J. J. Thomson, MacCullagh, Morton, Searle, Heaviside, Hall, and FitzGerald, of a search for a distinct relationship between momentum and the movement of charges connected to mass, through an interpenetrating "solid" ether, rather than the gaseous, dynamic, incompressible, ultra-fine ether of Tesla, existing in all space and ponderable matter, upon which electro-mechanical ponderomotive forces act. Once the equilibrium of the ether was "disturbed" by the moving charges, the 'displacement' could be rectified only by an equal and opposite reaction, which was the propulsion of the ship. Thomson had accepted the principle that the ether itself is the vehicle of mechanical momentum. The Hall effect had shown that an electromotive thrust (at close range) is produced along a third axis as a result of a current and magnetic field at right angles. Though Tesla alleged that this thrust could not be produced "in the free ether", but only in ponderable matter, the works of Heaviside, Searle, and Morton showed that the moving charges could either increase or decrease the normal ease of movement of a body, which inferred the feasibility of Tesla's theory.**

---

32[32.] J. Larmor, Phil. Trans. clxxxvi (1895), p. 697.

**Nikola Tesla's 1938 retrospective statement[33], that he had his Dynamic Theory of Gravity "all worked out" by 1894. Some available documentation of Tesla's work of 1891 or earlier shows that he was already experimentally ahead of the European field led by Wm. Thomson, J. J. Thomson, Searle, Morton, and Larmor—to whom he undoubtedly owed a debt. Though some of their statements were dated earlier, they lacked Tesla's essential hardware to verify their theories, or a complete valid theory.**

**In 1915, Tesla stated[34] that his (electro-mechanical) "...manless airship..." would travel "...300 miles a second..." (1.08 million mph), "...without propelling engine or wings, sent by electricity to any desired point on the globe...".**

**In 1940, Tesla stated[35] that he had already tested his four-part Teleforce system, which included "...a new method for producing a tremendous propelling force...", as used on his flying machine, in case anyone noticed.**

## CHAPTER V: TESLA AND THE GOOD OLD BOYS' CLUB

**When Tesla popped into the picture, the British "Good Old Boys' Club" had been debating ether theory for quite some time, and the upstart Tesla must have hurt the pride of their linemen, by making an end run to make a touchdown. After his 1892 lectures in London, they said nothing about him in their journals, never mentioned him. But he had showed them his hardware and how it worked. So they did mention "Tesla coils" and nothing else.**

**In 1847, Wm. Thomson, in discussing the motion of a magnetizable body in a non-uniform field of force, said a charged body attracts a body having a greater specific inductive capacity than that of the surrounding medium, and repels a body with a lower specific inductive capacitance, to afford the path of best conductance to the lines of force.**

**Thomson had also stated that an electrode immersed in a fluid insulating medium (apparently an analogy to a body in ether-filled**

---

33[33]. Nikola Tesla, Lecture before the Institute of Immigrant Welfare (May 12, 1938).

34[34]. Tesla Article, New York Times, (Dec. 8, 1915).

35[35]. Wm. L. Laurence, New York Times article (Sept. 22, 1940).

**space), at "...sufficiently high frequency", would cause a gravitation of gases all around toward the electrode, but that the general opinion (of he and his European colleagues) was that it was "out of the question" that such frequencies could be reached. While this was theory on Thomson's part, Tesla had already done the experiment. And Thomson's last opinion regarding frequency was soon to be disproved by Tesla, a close follower and admirer of Thomson's work.**

**In reiteration, the other Thomson—J. J. Thomson—had claimed to have mathematically developed the theory of moving tubes of force[36]. In his *Recent Researches in Electricity and Magnetism*, his hypothesis was that "...the aether is a storehouse of mechanical momentum," as stated earlier in this book.**

**Tesla's lecture before the A.I.E.E., Columbia College in 1891 was based on earlier experiments. He mentioned the "tubes of force" and disclosed some of his discoveries concerning the ether and momentum. His Feb. 1892 lecture[37] in London was made at a time when the Good Old Boys were still debating whether an electromagnetic action could occur in the free ether, at which time Tesla explained that he planned to run motors at a distance by wireless energy, with equipment he had already built, and to extract free energy from the environment. They must have been paralyzed by miscomprehension.**

**Four years later, Wm. Thomson stated[38] his "inclination to speculate" that "alterations of electrostatic force due to rapidly changing electrification" are propagated by "...condensational waves in the luminiferous aether." This statement seemed to indicate that Thomson was just beginning to take Tesla seriously. By "condensational waves" he apparently meant Tesla's mechanical pressure waves.**

**In his 1892 London lecture before the Good Old Boys at the Royal Academy, Tesla had stated that the 'required' frequencies—which Thomson had said were "out of the question" to be produced —were "...much lower than one is apt to estimate at first", and continued (in pertinent part, emphasis mine): "*We may cause the***

---

36[36.] J. J. Thomson, Supra.

37[37] Nikola Tesla, Lecture before the Institute of Electrical Engineers, London (Feb., 1892).

38[38] Bottomley, Nature liii (1896), p. 268.

*molecules of the gas to collide by the use of alternate electric impulses of high frequency, and so we may imitate the process of a flame; and from experiments with high frequencies which we are now able to obtain, I think the result is producible with impulses which are transmissible through a conductor." "...it appeared to me of great interest to demonstrate the rigidity of a gaseous column"..."with such low frequencies as, say 10,000 per second which I was able to obtain without difficulty from a specially constructed alternator." "...how must a gaseous medium behave under the influence of enormous electrostatic stresses which may be active in the interstellar space, and which may alternate with inconceivable rapidity?"*

**Tesla was already talking about outer space. In this respect, Tesla seemed also to address his Primary Solar Rays. His statements also show he was attempting to make up his mind as to the characteristics of the ether, such as its rigid or fluidic states, and under what circumstances it may change, and its static or dynamic nature, of high or low density, and so forth:** *"What determines the rigidity of a body? It must be the speed and amount of moving matter. In a gas the speed may be considerable, but the density is exceedingly small, in a liquid the speed would be likely to be small, though the density may be considerable; and in both cases, the inertia resistance asserts itself. A body might move with more or less freedom through the vibrating mass, but as a whole it would be rigid."*

**This statement reflects Tesla's earlier tests, as stated (emphasis mine) in his 1891 lecture before the A.I.E.E. at Columbia College in New York, repeated during his 1892 lectures in London,** *"...he illustrated how mechanical motions are produced by a varying electrostatic force acting through a gaseous medium".* **The tubes of force were said to be** *"...formed of rows of moving directed molecules"* **(in this sense, Tesla was referring to "ether molecules"). Tesla looked upon** *"...all electrostatic, current and magnetic phenomena as being due to electrostatic molecular forces."* **This statement clarified what Tesla meant by "molecules", and this is where Tesla parted ways with J. J. Thomson, who considered these phenomena to be of a purely electromagnetic nature, while Tesla regarded electrostatic phenomena to be more important.**[39]

39[39] Nikola Tesla, Lecture before the Franklin Institute, Philadelphia (Feb., 1893); Lecture before

**These lectures addressed how "solid bodies" can pass through a vibrating, incompressible mass of ether which, as a whole, is rigid. This is the essence of how the "inertia resistance" of the underlying ether framework can be summoned up by an electrified body which activates and rigidifies the ether with electrostatic impulses of "sufficiently high voltage and frequency". As the inertia resistance of the ether "asserts itself" the electrified body can be propelled through the ether.**

**Tesla's "specially constructed alternator" was in one embodiment a 32-inch diameter one, which, if similar to the type used on the saucer I saw in 1953, was probably driven by one of Tesla's bladeless turbines. In the 1890s, Tesla said the alternator had produced up to 10 amps and 30 kilocycles, like the one below:**

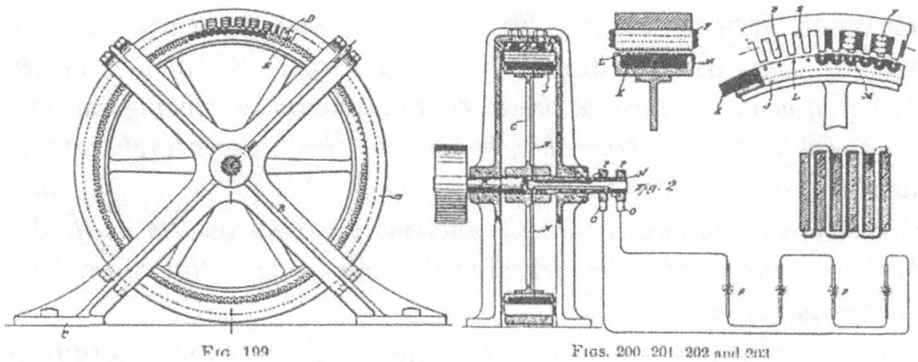

Fig. 199 Figs. 200, 201, 202 and 203

**The saucer which I saw in 1953 exhibited precessional characteristics which could have been caused by the use of such an alternator, the output and rpms of which under the circumstances could have been varied with the power level of the saucer. The high angle and low frequency of the precessions would be consistent with the use of a large diameter alternator, which was turned more slowly at hovering power, and increased in rpm for more power as the ship accelerated.**

**Since the alternator would have been attached to the airframe of the saucer, it could have caused the entire saucer to precess at a high angle and low frequency at low rpms. But this appeared to be a malfunction or perhaps a defect in the saucer at that time which had**

the National Electric Light Association, St. Louis (March, 1893).

not yet been corrected. A "double plate" gyrostabilizer idea which I mention in the following "bare-bones" saucer design would probably have corrected this problem, unless it was due to a turbine problem, with the turbine turning too slowly to maintain correct stability. On the other hand, the problem could have been caused by a malfunction of the main gyrostabilizer system in which case the rotating alternator's precessional characteristics predominated. This situation is shown in the last illustration in this book, the saucer which I saw in 1953.

The balanced forces holding the ship in mid-air would have been equivalent to holding it on "gimbals of air", so that it could have been precessing freely according to the speed of the rotating alternator's angular momentum and mass. This would have required very little force if the main gyro-stabilizer was malfunctioning, but points up the need for the better gyro-stabilization system which Tesla emphasized in his 1912-14 statements.

Tesla had anticipated the problem of how to counteract the tendency of his VTOL-type ship to rotate due to the torque of the propeller, alternator or turbine, by using two turbines or alternators, turning on parallel axes in the same direction or counter-rotated, as stated in his patent #1,655,114, Apparatus for Aerial Transportation, Jan. 3, 1928. In the case of a saucer however two-axis gyro-stabilization is necessary to stabilize the ship, one on horizontal axis to stop the spin of the ship, the other on vertical axis to keep the ship upright. And then a special gyro-stabilizer would be required to stop the wild gyrating of the ship caused by the alternator. But despite any possible malfunction of the main gyro-stabilizer, the high angle precession problem was solved when the alternator was spun up to high rpms as the ship departed. It just seemed such a big problem for the pilot and crew if they were trying to see where they were. I think by the way the ship made the two 90-degree turns to go about a hundred feet lower in altitude showed that they suddenly realized that they were over a town and wanted to diminish the visibility of the ship by the public by those maneuvers.

As the ship accelerated to full speed and power, its low precessional rate and high precessional angle became a mere high-frequency vibration, as the ship shot to infinity in three seconds (as I counted them off, "...thousand and one, thousand and two, thousand

and three…"). My calculation of the speed was based on my being able to see the ship for thirty miles. (I had previously miscalculated the speed by assuming that I could only see the ship for 7.5 miles, but I see aircraft coming over the Sandia Mountains thirty miles away all the time so this was wrong.) Since the ship covered ca. 30 miles in three seconds it was doing ca. 36,000 mph.

As the precession of the ship became the high-frequency vibration, this tended to blur the appearance of its edges, something which has probably made it more difficult to obtain sharp definitions of the profiles of saucers in photographs and videos. Coupled with this physical vibration may be the "Faraday effect"—the "magneto-optical effect" which tends to blur the outlines of objects subjected to intense electrostatic fields. The extension of the ship's electric field also extends its magnetic field, and causes a rotation of the optical plane, so in addition to visual effects of the high frequency precessional oscillations, the optical plane is actually rotated to create the weird magneto-optical effects so often reported, and becomes distorted in the minds of mystics, who think it is some sort of "time travel" or "inter-dimensional travel" effect, a "space-warping" or "wrapping around" of "time and space" by a "rotating body" as it moved through space ala Einstein, except saucers don't "revolve", as proven by my Peiltochterkompass and visual observations, and Einstein was full of baloney, as Tesla said.

The flying saucer may be powered by a Tesla alternator, powering a Tesla coil, with improvements which have probably come along in the meanwhile. The required currents could be conveyed through conductors by means of high voltage stepping switches or relays, to several coils facilitating the instantaneous control of a ship. Since an on-board power generator is usually required—unless the wireless Tesla transmission of power is used—the use of an alternator is convenient to run a spark gap, coil and condenser combination, since the necessary high frequency alternations can be easily stepped up to higher voltages by several closely linked "extra" coils distributed about the ship. That would make it easier to pipe medium high voltage current around to several coils as needed. The more modern development of very high K-factor dielectrics now makes it possible to more easily handle very high voltages and currents and greatly reduce the size of components, so that a high-output coil of several

**million volts can be made only a few inches in diameter, for example.**

**The master gyro-stabilizer for German ships, according to undocumented rumors from Germany, was purportedly constructed of two counter-rotating steel plates with pinion reversing gears between to create equal and opposite rotation. The precession and torque of one plate would cancel the precession and torque of the other plate to create the stability that Tesla sought, and may have been the "special" device which he referred to in 1914 in addition to the turbine, relied on for gyro-stability. For the high altitude flights that I observed in 1946-47, a pressurized cockpit would be required. Below is my illustration of the basic "bare bones" man-made UFO:**

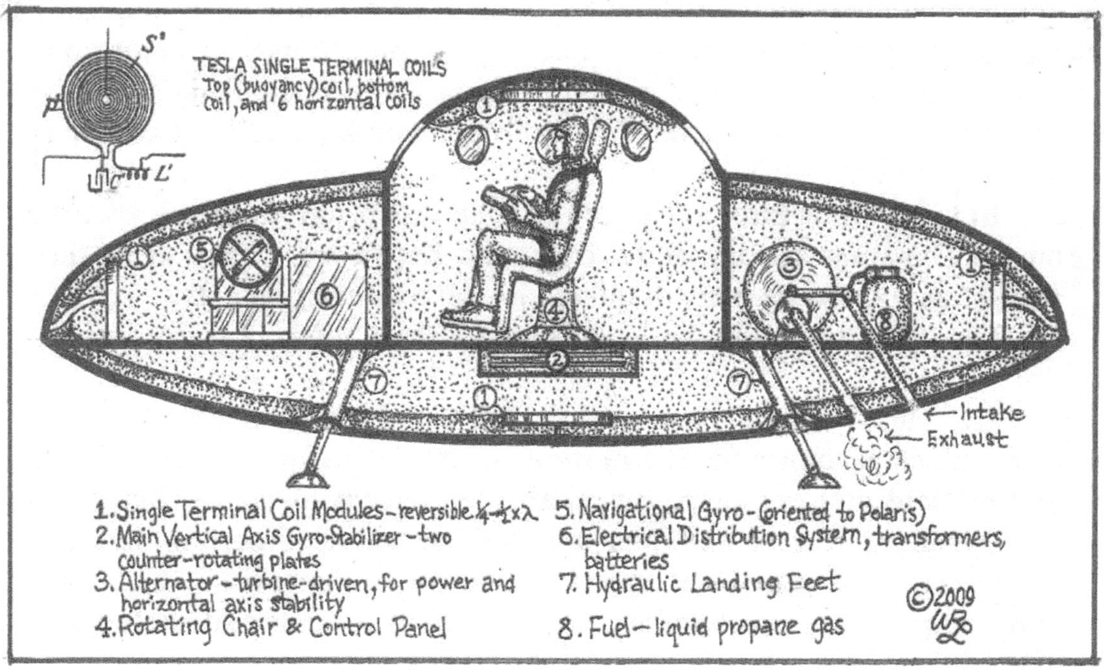


## TESLA'S DYNAMIC THEORY OF GRAVITY

**According to Tesla's 1938 Lecture before the Board of Immigrant Welfare[40], his Dynamic Theory of Gravity was one of two far reaching discoveries which he "...worked out in all details" in the years 1893 and 1894. Tesla never published his complete theory and**

---

40[40] Nikola Tesla (supra).

**the government now possesses all the papers related to it. The 1938 lecture was less than five years before his death.**

**More complete statements concerning these discoveries can only be gleaned from scattered and sparse sources, because the papers of Tesla are concealed in government vaults for "national security" reasons since 1945. When I specifically asked for these papers at the "National Security Research Center"—now the "Robert J. Oppenheimer Research Center"—at Los Alamos National Laboratory in 1979, though they admitted having the papers, I was denied access because they were classified and I didn't have a "badge" (clearance), and even though on that same day, I discovered the plans for the hydrogen bomb on an open shelf, and told a Harvard graduate student about it later in the day at a Santa Fe restaurant. The student went to Los Alamos, copied the plans and wrote an exposé at Harvard. I asked myself, "What could be in Tesla's papers in 1943 which is more secret than hydrogen bomb plans?" Of course I already knew the answer.**

**In his 1938 lecture, Tesla said that he was progressing with work on his Dynamic Theory of Gravity, and hoped to "...give it to the world" as soon as he had completed his secret developments.**

**The "two great discoveries" to which Tesla referred were:**

**1. The Dynamic Theory of Gravity - which assumed a field of force which accounts for the motions of bodies in space; assumption of this field of force dispenses with the concept of space curvature ala Einstein; the ether has an indispensable function in the phenomena (of universal gravity, inertia, momentum, and the movement of heavenly bodies, as well as all atomic and molecular matter); and,**

**2. Environmental Energy - the discovery of a new physical Truth: There is no energy in matter other than that received from the environment.**

**In the usual Tesla birthday announcement—on his 79th birthday (1935)—Tesla made a brief reference to the theory saying it applies to molecules and atoms as well as to the largest heavenly bodies, and to "...all matter in the universe in any phase of its existence from its**

**very formation to its ultimate disintegration."**

**Those imbued with relativist theory often refer to "pure energy" in some "form", but there is no such thing, since "energy" is an abstract "ability" which has no physical form as an entity unto itself. Who's to say what "form" is "pure" and what form is not?**

**My favorite philosopher, Ayn Rand, said "...In reality, there are no contradictions. Things are what they are irrespective as to whether we know it or not. Check your premises." If the term "energy" is only a convenient abstraction, then it does not exist in physical form, and really describes a process, the potential to perform work as a by-product of matter and electromagnetic radiation in perpetual motion, some of the force of which has been diverted through a path where it performs the desired work, as it goes on its way though the universe. Every change of form of form, matter or radiation involves the "work" which induces the change, or the "work" which is induced by the change. Without work there is no change, but all work is ultimately the product of the universe in perpetual, self-sustaining motion, as a rule and not an exception.**

**As for Tesla's theory, we have hints, such as, that the earth is the "star of human birth." In poetic expressions, he referred to scientific meanings in statements such as, that by using the "thunderbolt of Jove" (the Indo-European sky god), man "annihilates time and space", an allusion to the use of electrostatics ("thunderbolts") to travel so fast, that time and space are "annihilated". Where the government has stolen his papers, we must search for meaning elsewhere. In an article[41], Tesla outlined his Dynamic Theory of Gravity in poetic form (as paraphrased by me):**

1. ***That the luminiferous ether fills all space;***

2. ***That the ether is acted upon by the life-giving creative force;***

3. ***That the ether is thrown into "infinitesimal whirls" at near the speed of light, becoming ponderable matter;***

---

41[41] John J. O'Neill, Prodigal Genius (1944) pp. 251-52, referring to the Tesla article *Man's Greatest Achievement*.

**4.** ***That when the force subsides and motion ceases, matter reverts to the ether (similar to "atomic decay", only non-relativistic)***

**5.** ***That man can harness these processes, to:***

**-Precipitate matter from the ether**
**-Create whatever he wants with the matter and energy derived**
**-Alter the earth's size**
**-Control earth's seasons (weather control)**
**-Guide earth's path through the Universe, like a space ship**
**-Cause the collisions of planets to produce new suns and stars, heat, and light**
**-Originate and develop life in infinite forms**

**Tesla was referring to unlimited energy, derived from the environment. Several of his major free energy discoveries have been the exclusive stolen property of our Secret Government. The conversion of a weaker force to a stronger force—electro-mechanical propulsion—used to control the much weaker gravity force, would accomplish more work in the same amount of time, and therefore produce "over unity" results.**

**Some of Tesla's unusual conceptualization of the ether had been nonetheless expounded piecemeal, in his preceding 1890s lectures.[42] He later railed against the limited and erroneous theories of Maxwell, Hertz, Lorentz and Einstein.**

**Tesla believed that the ether is composed of "independent carriers immersed in an insulating fluid". A "carrier" is a conductor. The ether therefor is a conductor. It is unclear whether Tesla meant that the "insulative fluid" in which the conductive ether is immersed is the ether's own insulative fluid or the air, which Tesla often referred to as an insulator. When Tesla spoke of "the medium", he meant the ether, as stated here (emphasis mine):** ***"...But the upper strata of the air are conducting, and so, perhaps, is the medium in free space beyond the atmosphere, and these may contain an opposite charge."*** **(Nikola**

---

42[42] T. C. Martin, Inventions, Researches and Writings of Nicola Tesla, (1894), Chapter XXV - *Introduction - The Scope of the Tesla Lectures.*

Tesla, *High Frequency and High Potential Currents*, Chapter XXVIII, The Philadelphia and St. Louis Lecture. On Light and Other High Frequency Phenomena, February and March, 1893). We know today that the ionosphere is positively charged and that it is conductive. Meanwhile, Tesla said the ether in outer space may have a negative charge, and is also probably conductive.

Tesla's ether was neither the "tenuous" ether with the "solidity of the hardest steel" of Maxwell and Hertz, nor the half-hearted, gaseous ether of Lorentz. Though Tesla's ether was gaseous it consisted of "...carriers immersed in an insulating fluid", which filled all space, with properties which varied according to relative movement, the presence of mass, and the electric and magnetic conditions it is subjected to. It's incompressibility and other qualities meant it could be used to produce motion of ponderable bodies.

Tesla's ether was rigidified by rapidly-varying electrostatic force, and was thereby involved in gravitational effects, inertia, and momentum, especially in the space near earth, since, as explained by Tesla, the earth is "...like a charged metal ball moving through space", which creates enormous, rapidly-varying electrostatic forces which diminish in intensity with the square of the distance from earth, just like gravity. Since the direction of propagation radiates from the earth, the so-called force of gravity is toward the earth, and this apparently also applied to all other bodies in space under the theory.

Tesla commenced to complete his theory at the same approximate period of time that many of his experimental results and observations had been revealed in the three lectures, often illustrated with demonstrations using Tesla-invented equipment, as revealed in the following eight excerpts (the statements in quotes are Tesla's actual documented words) in pertinent part (emphasis mine):

1. "The most probable medium filling the space is one consisting of *independent carriers immersed in an insulating fluid.*"
2. "In his experiments he dwells first on some phenomena produced by *electrostatic force, which he considers in the light of modern theories to be the most important force in nature for us to investigate.*"
3. *"He illustrates how mechanical motions are produced by a varying*

*electrostatic force acting through a gaseous medium."*

**4. "One of the most interesting results arrived at in pursuing these experiments, is the** *demonstration of the fact that a gaseous medium upon which vibrations are impressed by rapid changes of electrostatic potential, is rigid."*

**5. "If through this medium enormous electrostatic stresses are assumed to act, which vary rapidly in intensity, it would allow the motion of a body through it, yet it would be rigid and elastic, although the fluid itself might be devoid of these properties."**

**6. "...on the assumption that the independent carriers are of any configuration such that the fluid resistance to motion in one direction is greater than in another, a stress of that nature would cause the carriers to arrange themselves in groups, since they would** *turn to each other their sides of the greatest electrical density,* **in which position the** *fluid resistance to approach would be smaller than to receding."*

**7. "If in a medium of the above characteristics a brush would be formed by a steady potential, an exchange of the carriers would go on continuously, and there would be less carriers per unit volume in the brush than in the space at some distance from the electrode, this corresponding to rarefaction."**

**8. "If the potentials were rapidly changing, the result would be very different; the higher the frequency of the pulses, the slower would be the exchange of carriers; finally, the motion of translation through measurable space would cease and, with a sufficiently high frequency and intensity of the stress, the carriers would be drawn towards the electrode, and compression would result."**

**In 5. above, Tesla states that an assumption exists that enormous, rapidly-varying electrostatic stresses exist in the natural environment,which rigidify the ether , yet allow the motion of a body through it. In 6., he assumes that the carriers can be induced to cause motion of a body in one direction. In 7., he mentions that a steady D.C. potential brush will cause a continuous rarefaction of the ether carriers in one direction to induce the pulling motion via the "veritable ropes of air" which Tesla mentioned in 1914. And in 8., he gives the conditions for compression by high frequency potentials which will stop the carriers from passing through the solid body, so**

**as to force the tubes of force from the opposite direction to dissolve in the body and impart momentum.**

**The eight excerpts are further reducible to the following four statements pertinent to his electro-mechanical propulsion technology:**

**1. Mechanical motions of solid bodies can be produced by varying electrostatic force acting through a gaseous medium (ether), which thereby becomes rigidified, yet allows solid bodies to pass through.**

**2. Under influence of stress in one direction, the carriers may group together, forming tubes of force, creating greater ease of movement in that direction.**

**3. When a D.C. brush is created by a steady potential in one direction, a continuous exchange of carriers is created corresponding to ether rarefaction, as the tubes of force are drawn into the conductor.**

**4. With a sufficiently high frequency (higher that that creating the brush discharges) and intensity of stress in the opposite direction, carrier exchange is blocked by ether compression, forcing the tubes of force entering the conductors of the ship from the opposite direction to dissolve and impart mechanical momentum.**

**The system, using the two kinds of potentials (D.C. and A.C.) is known as "p2", according to Tesla's (and that of the Germans under von Braun's direction) 1936-38 UFO project by that name at Los Alamos, New Mexico. My friend Peter van Dresser was the first person to tell me of von Braun's pre-war stay in New Mexico. Peter was distressed to the end of his life by the fact that, while he had invented the first successful liquid-fueled rocket engine, Dr. Robert Goddard had not only misappropriated Peter's invention, but had given it over to von Braun during his two-year stay in New Mexico. Peter considered Goddard a traitor and invention thief for this reason. Prior to Peter's invention, all the rockets of von Braun and Goddard exploded at an altitude of around 5,000-6,000 feet because the combustion chamber overheated. Peter wrapped the fuel lines around the chamber, cooling it and vaporizing the fuel for greater efficiency, making continuously extended flights possible.**

**In reiteration, the steady D.C. Electrostatic potential of the brush creates the required exchange of carriers, 'rarefying' (stretching) the elastic, rigidified medium (composed of ether carriers in an insulating fluid) in advance of the ship, as the higher frequency A.C. to the rear compresses them, blocking exchange in that direction, dissolving the tubes of force from the opposite direction and creating instant momentum, normal to the surface.**

**In 1884, John Henry Poynting's theorem had been that the flux of energy at any place is represented by the vector product of the electric and magnetic forces, multiplied by $C/4\pi$.[43] This implied that forces in a conductor could be transformed there into other forms (such as the transformation in the conductors of the tubes of force into momentum). In 1893, J. J. Thomson stated practically the same thing, saying "...the aether is itself the vehicle of mechanical momentum, of amount $(1/4\pi C)$ $(D{\cdot}B)$ per unit volume.[44]**

**(Using e.-s. Units for D and E**
**And e.-m. Units for B and H.)**
**E = electrical force**
**D = electrical displacement**
**H = magnetic force**
**B = magnetic induction**

**Heinrich Hertz's theory[45] was that two systems of varying current should exert a ponderomotive force on each other due to the variations. Tesla's disagreement was apparently based on the fact that he proved that the "ponderomotive force" is due not to mere "varying currents", but to rarefaction and compression of the ether carriers, respectively, produced by *different kinds of currents* (D.C., rapidly varying electrostatic and A.C.) Moreover, Tesla's method for accessing the mechanical momentum of the ether was from a completely different and simpler perspective, the creation of (longitudinal) sound waves in the ether in the direction of desired propulsion. This also appears to be the method which allows the ships to eliminate drag and sonic booms.**

---

43[43] John Henry Poynting, Phil. Trans. clxxv (1884), p. 343.

44[44] J. J. Thomson, Recent Researches in Elect. And Mag. (1893), p. 13.

45[45] Heinrich Hertz, Ann. d. Phys. Xxxi (1887), p. 421; *Hertz's Electric Waves*, translated by D. E. Jones, p. 29.

**J. J. Thomson[46] had extensively developed the theory of the moving tubes of force, both magnetic and electric, saying that the magnetic effect was a secondary one created by the movement of electric tubes, and assumed:**

- ➔ **That tubes exist everywhere in space, either in closed circuits or terminating on atoms;**
- ➔ **that electric force becomes perceivable only when electric tubes have a greater tendency to lie in one direction;**
- ➔ **that in a steady magnetic field, positive and negative tubes may move in opposite directions with equal velocity;**
- ➔ **that a beam of light is a group of electric tubes moving at C at right angles to their length.**

**Tesla later said his "dirigible torpedo" ("steerable" torpedo) would fly at a maximum 300 miles per second, perhaps since its forward velocity would be some discreet fraction of C. Thomson's 1893 publishing on this subject followed Tesla's 1891 lectures before Franklin Institute and Columbia College in America, and his lecture before the Royal Society in London, and appeared to shed light on Tesla's work, stating:**

- ➔ *That a ponderomotive force is exerted on a conductor carrying electric current, consisting of a transfer of mechanical momentum from the agent which exerts the force to the body which experiences it;*
- ➔ *that, if moving tubes entering a conductor are dissolved in it, mechanical momentum is given to the conductor;*
- ➔ *that such momentum must be at right angles to the tube and to the magnetic induction;*
- ➔ *that momentum stored in a unit volume of the field is proportional to the vector product of electric and magnetic vectors.*

**Thomson's** ***Electromagnetic Momentum*** **hypothesis was later developed by H. Poincaré[47] and by M. Abraham[48].**

---

46[46] J. J. Thomson (supra).

47[47] H. Poincaré, Archives Né erl. (2) v (1900), p. 252.

48[48] M. Abraham, Gött., Nach., (1902), p. 20.

**But we can see a distinct difference between Tesla's and Thomson's theories in that Tesla said nothing about the electric and magnetic fields, relying solely on the effects of the combination of a rapidly-varying electrostatic D.C. (rarefying) potential and a higher frequency rapidly-varying electrostatic A.C. (compressing) potential, both produced by a Tesla coil with different tuning.**

**By 1910, it was said[49] (by Sir. Edmund Whittaker) that the consequences of these pronouncements left three alternatives:**

**(1) Modify the theory to reduce to zero the resultant force on an element of free aether (as with Maxwell, Hertz, and Einstein);**

**(2) assume the force sets aether in motion (as with Helmholtz);**

**(3) accept the principle that aether is the vehicle of mechanical momentum of amount [D·B] per unit volume (as with Poynting and J. J. Thomson).**

**Whittaker's greatest error was in omitting mentioning Tesla's theory entirely, even though Tesla's experiments had verified it, in front of the esteemed members of the "Royal Academy". It seems highly improbable that they simply failed to comprehend Tesla. Tesla however said he had not "completely" worked out his Dynamic Theory of Gravity until 1893 and 1894, but never published it, so how could it be discussed? In reality, the "three (later) alternatives" were moot, and a new law existed, that of Tesla. So we see that the suppression of Tesla's theory was partially a conspiracy and partially his own fault.**

## Tesla's Secrecy

**Due to his pacifist sympathies, Tesla originally contemplated giving his electric flying machine to the Geneva Convention or League of Nations, for use in 'policing the world' to prevent war. Later disillusioned after WW I with the collapse of the League, he said he'd "underestimated man's combative capacity."[50]**

**In 1919, Tesla's reason for increased secrecy emerged[51], while**

49[49] Sir Edmund Whittaker, A History of the Theories of the Aether and Electricity, (1910) Edinborough.

50[50] Nikola Tesla, New York Times, article (July 10. 1934).

51[51] Tesla Interview with Frederick M. Kerby, Resolution Magazine, (1919).

**discussing his plans for a "three-hour" airplane between New York and London:** ***"...we have here the appalling prospect of a war between nations at a distance of thousands of miles, with weapons so destructive and demoralizing that the world could not endure them. That is why there must be no more war."*** **With the U. S. government's spurning of his defense suggestions and technology, Tesla's only recourse was to withhold his Dynamic Theory of Gravity, and to dissuade discovery in that direction. But he had already sold his secrets to the Germans in 1914.**

**In 1929[52] Tesla ridiculed Heinrich Hertz's 1887-89 experiments purportedly proving the Maxwellian "structureless" ether filling all space,** ***"of inconceivable tenuity yet solid and possessed of rigidity incomparably greater than the hardest steel."*** **Tesla's arguments were to the contrary, saying he had always believed in a "gaseous" ether in which he had observed (longitudinal) waves more akin to sound waves. He recounted how he had developed a "new form of vacuum tube" in 1896,** ***"...capable of being charged to any desired potential, and operated it with effective pressures of about 4,000,000 volts."*** **He described how blue coronal discharges about the bulb when in use, verified the existence of** ***"particles smaller than air"*****, and a gas so light that an earth-sized volume would weight only 1/20 pound. He further said that so-called "electromagnetic waves" are actually longitudinal---not transverse---sound waves moving at the velocity of light through this medium.**

**In 1932[53] Tesla mentioned using his special tube to investigate cosmic rays, saying that when its emanations were impinged upon a target material, radioactive emissions resulted, and that radioactive bodies were simply "targets" continuously bombarded by** ***"infinetesimal bullets projected from all parts of the universe"*****, without which** ***"all radioactivity would cease"*****. His name for these "bullets" was "Primary Solar Rays".**

**Later in 1932[54], Tesla said Einstein's theory regarding changing matter into force, and force into matter, was "absurd". He compared this to the difference between body and mind, saying force is a**

---

52[52] Nikola Tesla interview, New York Herald Tribune, (Sept. 22, 1929), pp. 1, 29.

53[53] Letter, New York Times, (Feb. 6, 1932), p. 16, col. 8.

54[54] *Nikola Tesla Papers*, Rare Books and Manuscript Library, Columbia University (Apr. 15, 1932).

***"...function of matter"*, and that, just as a mind could not exist without a body, *"...without matter, there can be no force."***

**Again in 1932[55], Tesla derided the Maxwellian/Hertzian ether, while saying that higher frequency waves *"...follow the curvature of the earth and bend around obstacles"*, yet in Apr. 8, 1934[56], Tesla said that short waves for "power purposes" of the 'wireless art', were inappropriate, and that power will travel in "long waves". All this in retrospect to deaf ears since no one in the public's knowledge uses wireless technology to transmit energy today, and no one since Tesla has been able to do so because they haven't paid close attention to all that Tesla said (that we in the public know of).**

**With so many "experts" today claiming to know Tesla technology such as those who say that Tesla used "quaternion equations" ("curved space-time" and "fourth-dimensional" math, ala relativism), I will settle once and for all the question of authenticity of these "experts" on Tesla. Lest there be any misunderstanding or confusion about Tesla's beliefs, here is a 1932 newspaper quote under the headline "Pioneer Radio Engineer Gives Views on power:**

***"The assumption of the Maxwellian ether was thought necessary to explain the propagation of light by transverse vibrations, which can only occur in a solid. So fascinating was this theory that even at present it has many supporters, despite the manifest impossibility of a medium, perfectly mobile and tenuous to a degree inconceivable, and yet extremely rigid, like steel. As a result some illusionary ideas have been formed and various phenomena erroneously interpreted. The so-called Hertz waves are still considered a reality proving that light is electrical in its nature, and also that the ether is capable of transmitting transverse vibrations of frequencies however low. This view has become untenable since I showed that the universal medium is a gaseous body in which only longitudinal pulses can be propagated, involving alternating compressions and expansions similar to those produced by sound waves in the air. Thus, a wireless transmitter does not emit Hertz waves which are a myth, but sound waves in the ether, behaving in every respect like those in the air, except that, owing to the great***

---

55[55] Nikola Tesla, New York Herald Tribune, (Sept. 11, 1932).

56[56] Nikola Tesla letter, New York Times, (Apr. 8, 1934).

*elastic force and extremely small density of the medium, their speed is that of light. general knowledge of this subject seems very limited, I may state, that even waves only one or two millimeters long, which I produced thirty-three years ago, provided that they carry sufficient energy, can be transmitted around the globe. This is not so much due to refraction and reflection as to the properties of a gaseous medium and certain peculiar action which I shall explain some time in the future. At present it may be sufficient to call attention to an important fact in this connection, namely, that this bending of the beam projected from reflector does not affect in the least its behavior in other respects.*

*It might be inferred that I am alluding to the curvature of space supposed to exist according to the teachings of relativity, but nothing could be further from my mind. I hold that space cannot be curved, for the simple reason that it can have no properties. It might as well be said that God has properties. He has not, but only attributes and these are of our own making. Of properties we can only speak when dealing with matter filling the space. To say that in the presence of large bodies space becomes curved, is equivalent to stating that something can act upon nothing. I, for one, refuse to subscribe to such a view."*

**Anyone who says otherwise after reading this quote is definitely a liar or someone who has serious reading comprehension problems and shouldn't be taken seriously. You may agree or disagree with it but you can't deny that this is what the man said and meant.**

## THE NATURE OF ELECTRICITY

**What were the old ether physicists referring to when they attempted to describe "an incompressible, perfect fluid"? What would a "perfect fluid" do? It would be able to "wet" and soak into everything it came into contact with, such as protons, and neutrons, and could flow everywhere often without resistance (except in the circumstances already described). This fluid, the ether, because of its ultra-fineness, could flow everywhere. Grosser matter on the other hand, experienced resistance, depending on the circumstances. Another fluid—electricity—could flow in certain places, and wet only certain things, but often met resistance.**

**In his 1891 A.I.E.E. lecture at Columbia College, Tesla said (in pertinent part, emphasis mine):** *"What is electricity, and what is magnetism? "...We are now confident that electric and magnetic phenomena are attributable to the ether, and we are perhaps justified in saying that the effects of static electricity are effects of ether in motion." "...we may speak of electricity or of an electric condition, state or effect." "...we must distinguish two such effects, opposite in character neutralizing each other." "...for in a medium of the properties of the ether, we cannot possibly exert a strain, or produce a displacement or motion of any kind, without causing in the surrounding medium an equivalent and opposite effect." "...its condition determines the positive and negative character." "We know that it acts like an incompressible fluid;" "...the electro-magnetic theory of light and all facts observed teach us that electric and ether phenomena are identical." "The puzzling behavior of the ether as a solid to waves of light and heat, and as a fluid to the motion of bodies through it, is certainly explained in the most natural and satisfactory manner by assuming it to be in motion, as Sir William Thomson has suggested." "Nor can anyone prove that there are transverse ether waves emitted from an alternate current machine; to such slow disturbances, the ether, if at rest, may behave as a true fluid."*

**In his statements, Tesla was balancing the various arguments in preparation for his decision:** *"...Electricity, therefore, cannot be called ether in the broad sense of the term; but nothing would seem to stand in the way of calling electricity ether associated with matter, or bound ether; or, in other words, that the so-called static charge of the molecule is ether associated in some way with the molecule."*

*"...It cannot differ in density, ether being incompressible; it must, therefore, be under some strain or in motion, and the latter is the most probable."*

**Tesla therefore believed in 1891 in an ether which was in motion relative to earth. I think this belief was modulated in Tesla's 1899 Colorado Springs Experiments due to his discovery that the earth emits rapidly-varying electrostatic forces which rigidify the ether in the earth's electric field and create the "carry along" ether, while the tubes of force induced by this are in relative motion.**

**The thing which Tesla had realized, was that ether possesses electric charges which are deposited on molecules. His "dynamic" ether belief is consistent with the idea that everything is in motion.**

**The importance of cosmic motion to the effects of electrostatic**

**charges was apparently brought up by Tesla in his lecture:** ***"About fifteen years ago, Prof. Rowland demonstrated a most interesting and important fact, namely that a static charge carried around produces the effects of an electric current." "...and conceiving the electrostatically charged molecule in motion, this experimental fact gives us a fair idea of magnetism. We can conceive lines or tubes of force which physically exist, being formed of rows of directed moving molecules; we can see that these lines must be closed, that they must tend to shorten and expand, etc. It likewise explains in a reasonable way, the most puzzling phenomenon of all, permanent magnetism, and, in general, has all the beauties of the Ampere theory without possessing the vital defect of the same, namely, the assumption of molecular currents. Without enlarging further upon the subject, I would say, that I look upon all electrostatic, current and magnetic phenomena as being due to electrostatic molecular forces."***

**In these statements, Tesla showed he was aware that any "stationary" locale on earth is actually in fantastic motion ("70,000 mph"). The electrostatic charges "carried around" are currents between atoms and the ether, which produce magnetism. The phenomena of 'permanent magnetism' or 'cosmically induced' magnetism are apparently due to electrostatic charges 'carried around' by cosmic molecular motion, in the universal ether field.**

**Since no one can hold atoms, electrons or molecules perfectly still—because they are in fantastic motion—all atoms and molecules carry currents producing magnetic fields. Since a magnetic field is the product of a current, no one can produce a magnetic field without electricity, moving through or along a conductor, or as electrostatic charges in local or cosmic motion.**

## CONCLUSION

**Since these phenomena are electro-mechanical in nature, they are synthesized by recreation of the electromagnetic conditions which a body would exhibit due to a particular kinetic state. Just as sound waves of low frequency can penetrate a body, waves of higher frequencies can cause the ether to "assert its inertia resistance."**

**Tesla stated that the steady "brush" discharge causes a steady exchange of carriers, but emphasized the necessity of the "rapidly-**

**varying electrostatic" impulses to create sound waves in the ether to cause it to assert its "inertia resistance" and form the tubes of force. Here is where Tesla said the best of electrical engineers would not believe it, that the rapidly-varying electrostatic potentials create the steady D.C. potential which causes a steady exchange of carriers, which brings in the tubes of force to propel his ship. Because the D.C. potential is undisturbed yet continuous and rapidly-varying, it is nonetheless "steady".**

**The technology involves high voltage ("D.C.") electrostatic 'brush' currents on top of the ship and which rarefies the elastic ether and creates the bouyancy of the ship, and in front which stretches ("rarefies") the elastic ether and pulls the ship forward, while a higher frequency A.C. current compresses the ether and blocks ether from passing through the rear of the ship. These currents force the tubes of force to dissolve in the conductors of the ship and impart momentum. Since the strength of the electrical attractive force is 2 x $10^{39}$ power times stronger than the gravitational attractive force, that much more work can theoretically be done in the same amount of time, using the same "energy". And that is how the ship I saw in 1953 accelerated to ca. 36,000 mph in three seconds.**

**Since "seeing is believing", and I have seen, I believe. The behavior of man-made flying saucers proves the existence of these "free energy" inventions of Nikola Tesla, which show that he was right in his opposition to Relativism, and that the prevalent theories taught in the scientific institutions of the world today are false.**

**Consistent with the idea that "dynamite comes in small packages", this small book, along with my prior book, *Space Aliens From the Pentagon*, along with my third edition of that book, *Pentagon Aliens*, have all initiated the rediscovery, reconstruction and publication of Nikola Tesla's *Dynamic Theory of Gravity* and related electro-mechanical propulsion technology, the greatest invention of mankind.**

**The deep black Secret Government by whatever name, be it "illuminati", "cabal" or whatever —a fraudulently concealed, unconstitutional, corporate-state entity—has heretofore controlled electro-mechanical propulsion technology by concealing it along with other advanced free energy technology, on behalf of international,**

coercive, corporate-state monopolists. Electro-mechanically propeled ships are concealed through "special effects", and "stealth" technology (also based on Tesla inventions)—as they fly over and hover in the sky, and by "psycho-political" means—the dissemination of false "alien" and "extraterrestrial" origins propaganda—through "UFOlogy" groups led by covert government agents. While official government spokesmen deny the existence of flying saucers, covert government agents, posing as "skeptics", "UFOlogists", and "paranormalists", engage the public's attention in a phony debate. The "skeptics" ridicule the UFOlogists and paranormalists, lumping them together with rational witnesses, as if "...to see a flying saucer" is as ridiculous as "...to see a ghost or alien". This phony debate based on false logic fits nicely into the overall cover-up system, designed to conceal advanced human technology, not "alien" technology.

The grip of this secret socio-economic dictatorship depends on coercively extorted income, by forcing us to buy archaic fuels and power—which funds its control of our communications, political and monetary decisions, enforcement of unjust laws, regulations and procedures expanding its powers, while limiting or excluding our individual human rights, and denying our access to information, materials, and technology which is ours because we paid for it with our taxes. Our access to the truth is necessary for our fully-informed consent, independence, and survival as a free people.

Despite the confusion, concealment, BIG LIES, and judicial and socio-economic abuse by the corporate state, we can resurrect the TRUTH, determine the correct technology, circumvent obstructions, and use our creative ingenuity to build these ships for ourselves, to free our society from the Secret Fascist Corpocracy.

There is much confusion created by government agents or dupes regarding UFO propulsion technology. Do not be fooled. You will find no "patents" on this technology because it is secret, "classified" information by all the major governments of the world. Anyone who shows you a "patent" purporting to represent this technology is either a liar or a fool, and could only show you "how NOT to build a UFO". The same goes for anyone who jabbers idiotic nonsense about "space aliens". These craft are exclusively man-made.

**Below is a post from the Electrical Review [57], entitled, which also appeared in The Sun, N.Y., Nov. 21, 1898, and which contains some pertinent information relating to when Tesla began to develop his flying machine, more particularly what he called his "dirigible torpedo":**

*"Referring to my latest invention, I wish to bring out a point which has been overlooked. I arrived, as has been stated, at the idea through entirely abstract speculations on the human organism, which I conceived to be a self-propelling machine, the motions of which are governed by impressions received through the eye. Endeavoring to construct a mechanical model resembling in its essential, material features the human body, I was led to combine a controlling device, or organ sensitive to certain waves, with a body provided with propelling and directing mechanism, and the rest naturally followed. Originally the idea interested me only from the scientific point of view, but soon I saw that I had made a departure which sooner or later must produce a profound change in things and conditions presently existing. I hope this change will be for the good only, for, if it were otherwise, I wish that I had never made the invention. The future may or may not bear out my present convictions, but I can not refrain from saying that it is difficult for me to see at present how, with such a principle brought to great perfection, as it undoubtedly will be in the course of time, guns can maintain themselves as weapons. We shall be able, by availing ourselves of this advance, to send a projectile at much greater distance, it will not be limited in any way by weight or amount of explosive charge, we shall be able to submerge it at command, to arrest it in its flight, and call it back, and to send it out again and explode it at will, and more than this, it will never make a miss, since all chance in this regard, if hitting the object of attack were at all required, is eliminated. But the chief feature of such a weapon is still to be told; namely, it may be made to respond only to a certain note or tune, it may be endowed with selective power. Directly such an arm is produced, it becomes almost impossible to meet it with a corresponding development. It is this feature, perhaps, more than in its power of destruction, that its tendency to arrest the development of arms and to stop warfare will reside. With renewed thanks, I remain,*

*Very truly, yours,*

*N. Tesla"*

---

57[57.] Nikola Tesla, *Electrical Review* (Nov. 19, 1898) New York.

**Tesla didn't want to give the whole ball game away on his propulsion system, and was secretive about it, which is why he never published his completed *Dynamic Theory of Gravity.* By habit, if Tesla disagreed with a theory, he usually wasted little time in refuting it. In scenting out Tesla's secret ether theory that Tesla relied upon for his propulsion system, I compared it with a theory which he did not even mention, J. J. Thomson's *Electromagnetic Momentum* theory. While Tesla had in fact gotten into rather spirited exchanges with Thomson regarding electrostatic versus electrodynamic excitation of luminosity of gases in a tube at a distance, that was the most that Tesla was willing to risk. he never said a word directly regarding Thomson's Electromagnetic Momentum theory, which was wrong because it was based on the 'solid ether' and electromagnetic waves theory of Maxwell and Hertz.**

**The absence of a specific refutation or even a mention of it is a conspicuous omission which points to Tesla's thinking on this subject once one realizes that this is a pattern seldom if ever broken by Tesla. What this means is that, regarding Thomson's Electromagnetic Momentum theory, Tesla could not criticize Thomson's theory without revealing his own ether propulsion theory. He had already discussed his use of electrostatically-induced mechanical waves rather than the supposed electrodynamics of Thomson. Not only that, but Tesla could not even mention the theory specifically lest he draw attention to his own theory and disclose his most important secret.**

**In a similar vein however Tesla had earlier stated (Electric Discharges in Vacuum Tubes, *The Electrical Engineer*, N.Y., July 1, 1891) *"...the electrostatic effects were of far greater importance..."* and *"In using the term electrostatic I had reference rather to the nature of the action than to a stationary condition, which is the usual acceptance of the term."* To this Thomson agreed, saying *"As a matter of fact, it is very much easier to get the luminosity when these electrostatic effects are operative than when they are not."* (Note by Prof. J. J. Thomson in the London *Electrician*, July 24, 1891).**

**There is in my opinion, based on all my studies, only one kind of UFO propulsion technology and it is electro-mechanical, invented by Nikola Tesla. From my observations of the ship I saw at close range**

**in broad daylight in 1953, there were no intake or exhaust ports, no jets or rockets, no "plasma" repulsion system, etc. The ship was apparently sustained and moved without these, the only external evidence being high voltage electrostatic corona. When you have a ship which can fly at 36,000 miles per hour in 1953 you cannot want for more.**

**While Thomson's theory proposed a Maxwellian three-axis system which used electric and magnetic force on two of the axes to purportedly create momentum on the third axis, Tesla used "p2", a combination of electrostatic brush discharge in the ether projected in the direction of travel and high frequency A.C. in the opposite direction to block the ether and that was it. Tesla proved that Maxwell was wrong regarding electromagnetic waves in free space and rejected those theories, which were based on Hertz's erroneous theory of a solid ether.**

**Some of you reading this may have thought there was a contradiction between Tesla's statement that his flying machine was propelled by "mechanical" waves to propel his ship with a "pulling" action (ropes pull), then referred to the "steady D.C. brush discharge" as a means of creating the continuous exchange of carriers which imparts "mechanical" momentum to the ship. Both these means are combined, and Tesla also referred to "environmental" energy in free space, which can be harvested.**

**To put it into a nutshell, the rapidly-varying electrostatic potentials tuned to one-quarter wavelength produces the steady D.C. brush discharge which rarefies (stretches) the ether, while the rapidly-varying electrostatic potentials tuned to one-half or full (resonant) wavelength produces the reaction which compresses the ether. The sound waves in the ether are also used for transmitting wireless energy. A quarter wavelength tuning is for transmitting, while the half or full wavelength tuning for receiving, but at high frequencies has the effect of blocking the ether by compression.**

**Tesla's Dynamic Theory of Gravity and method of mechanical space propulsion brought a cosmic crowning achievement to the works of his esteemed predecessors who did so much of the ground work, but the public has been denied this achievement.**

# CHAPTER VI: FREE ENERGY MASSACRE;
## The Atomic Hydrogen Process
## ©1996, Wm. R. Lyne

**What better way to end a book on "Occult Ether Physics" than with a chapter on another concealed "free energy" process, which appears to extract energy from the environment—using similar techniques on the same substance—the ether—that makes Tesla's electro-mechanical propulsion possible?**

**In *Space Aliens From the Pentagon* and *Pentagon Aliens* I specified a free energy process in which helium is made to produce 460,000 calories/gram-atom by simple spark discharge, which is based on data from numerous generally available texts and scientific encyclopedias. This is the highest energy output I know of from a noble gas, and the same gas can be used *ad infinitum*, the only input energy being that required from spark discharges. Since helium is inappropriate for the average "Joe" and since the output is probably more than you could handle, I will more appropriately show you how to perform another free energy process, using that most easily accessible gas, hydrogen. This chapter will also give you the explanation for the heat produced in so-called "cold fusion" and a few other free energy processes, all closely related to the atomic hydrogen process and occult ether physics.**

**Over the years, the "Relativistic Quantum Mechanics" (hence "RQMs") concealed every process or reaction which could show the invalidity of their theories, "...in the interest of national security". Their false theories actually became one of the most important means by which, from the standpoint of the controlling corporate fascists, all 'dangerously efficient' (i.e., "disruptive") technology which would destroy their iron grip on us, could be concealed. According to deceitful relativist quantum mechanics, all energy coming out of a reaction must go into it only from a source recognized by them, and only in the discrete "quanta" permitted by Planck's earlier quantum theory. The basic idea of Planck's original quantum mechanical theory, was that energy levels of atoms must change in whole number integers (1, 2, 3, etc.) called "quanta", which are based on the basic charge of a single electron or multiples thereof, which assumes only certain assigned energy levels in different atoms and molecules, and**

is used to define the magnitude of all of the smallest, intermediate, and largest possible energy changes of atoms and molecules. All atomic energy changes must conform to the permissible whole number equivalents or multiples of these quanta. Accordingly, energy absorbed or emitted by an atom must fit this stringent criterion.

What, then, would you say about a process which completely destroys this neat plan, which causes atoms to emit more energy that that which was put into them?

Despite what I consider to be obvious attempts to conceal basic physical data conflicting with this principle—as defined by the RQMs and incorporated into their theory, as if it were exclusively their theory and not that published earlier by Planck—the process I am about to discuss with you has glaring inconsistencies in it, as described by the Relativists, in particular Irving Langmuir. Langmuir was wrong and my research, invention and experiments and tests based on it prove him wrong.

It is said by the RQMs that the process of dissociating diatomic hydrogen molecules into separate atoms as atomic hydrogen, requires the absorption of 109,000 cal./gram molecule ("1.9 k.cal./gram molecule"), and that that is all the energy you can possibly get out of it. Yet, in taking the basic hydrogen atom and considering all its discrete 'quanta'—whether in the atomic or molecular state—it appears to me that there is no state of the hydrogen atom or molecule in which such energy could be "stored", even when using the RQMs' theories,, especially in view of their irrational statement that the energy is still possessed by the molecule after it has already been released! It is their contention that an independent hydrogen atom can only exist as such because it has absorbed an abnormally large amount of heat, despite the fact that by their own criteria the atom has no mechanism identifiable by them by which it could do so.

When two such hydrogen atoms come together with the presence of sufficient electrostatic initiation energy, the great amount of heat said by the Relativists to be "stored" in the hydrogen atom is liberated. The RQMs never directly measured the amount of dissociation energy necessary to bring the hydrogen atom from the diatomic to the atomic state, because they didn't know how to do so, but because their theory said that this reaction is reversible, they said the dissociation energy was the same as the energy released on

**recombination of hydrogen atoms. Then to make matters worse, they defined the atomic hydrogen (ground) state as if it were an excited state, and the molecular state as if it were the ground state.**

**Realistically, the atomic hydrogen reaction can only be satisfactorily explained by reference to, construction of, or re-construction of, an ether theory and involving Tesla's Primary Solar Rays (one form of "environmental energy"). While it may be arguable that the electrostatic "binding energy" between the two atoms of the molecule somehow 'includes' this energy in some undefined and mysterious way, this argument actually supports an ether theory, since the binding energy (from somewhere) must somehow be exchanged with what is said to be the heat energy which is released when the molecule forms, according to their theory, consistent with the equal-and-opposite-reaction rule. Otherwise, one is expected to believe a kind of backwards Orwellian *doublethink* that the RQMs. This kind of latitude is permitted only for them, while those who disagree are held to a higher standard.**

**The doublethink (contradictory) features of Relativism apparently stem from the fact that Albert Einstein was a Kaballist-Existentialist—whose epistemology was a Platonist one. In philosophy, there are two basic epistemologies—theories of the origin, nature and limits of human knowledge—representing the two basic categories into which the epistemologies of all philosophies must fall, and which are supposed to explain how one knows something according to a particular philosophy...by what *means* they know it...if in fact, they *do* "know" it. Since he who asserts something has the burden of proving it, rather than one who disbelieves it, the means by which it is purportedly known may come into question. No one is legally, morally, or "spiritually' obligated to believe anything, without reason, or a valid means of knowing it to be "true". All epistemologies are classified as either *Platonist* or *Aristotelian*.**

**The "Platonist Epistemology" is so named because of the theory illustrated by the ancient Greek philosopher Plato's cave analogy, in his book, *The Republic*. In this analogy, an observer inside a cave (us), facing the inner wall at night, sees only the shadows cast upon the wall by such things as a passing caravan, projected by a fire burning**

beyond a road along which the caravan is passing. According to Plato, man cannot know reality, but sees only a 'shadow' of reality, because there is a 'higher world' of 'perfect forms', which exist only 'in heaven'. It therefore becomes obvious how this epistemology follows religious thinking, and that many religions have been influenced by it as somehow adding 'scientific credence' to their religious ideas, which are based on faith rather than reason. The same reasoning has been used by scientific, philosophical, economic, political, and 'spiritual' con-artists for centuries, and every totalitarian philosophy ever known has relied upon this fallacious 'reasoning' which is identical to the Platonist epistemology, because it relieves its advocates of showing reason and logic for what they assert. One wonders how the viewer in Plato's cave would draw the line between 'knowing' a shadow, and a "shadow of a shadow", since a shadow might also have a 'higher form' under this theory.

The "Aristotelian Epistemology" in the meanwhile—based on the ancient Greek philosopher Aristotle's theory of knowledge—holds that the senses are valid, scientific tools of cognition, which we can rely upon to detect the facts of perceivable reality, and that we can, by relying upon the facts so perceived, know reality—and the nature of reality—and based on that, determine other, perhaps unapparent facts of reality.

As can be seen from a comparison of these two epistemologies and their implications, the philosophies and endeavors based on the Platonist epistemology are mystical, nebulous and/or totalitarian because its advocates require you to believe things based on faith and not reason, while the Aristotelians are more rational and objective, especially when applied to the natural sciences. It can also be seen that the use of a Platonist Epistemology by a physicist should be considered unforgivable, while Einstein's theory, built upon this fatally flawed epistemology, has been popularized by the communication monopoly controlled by the power élite, because it makes a potentially inquisitive and questioning society dumber and easier to control. Einstein's image is continually built up as if that of an ultimate and infallible creature who should be regarded as a 'god'. This propaganda not only protects the false theory but also aids in concealment of valid theory and advanced technology which would end the reign of a world-wide power structure of corporate-statists,

**who think society already has all the technology it 'needs', else the élite's 'grip' on the world, held in place by their archaic technology which exploits mineral, industrial, and banking interests owned by them, might be torn loose. Although Einstein eventually admitted in the 1920s that 'there has to be an ether', he never followed up on that change in his ideas with anything concrete, although his friend Paul Dirac did in a paper he sent to Einstein arguing for an ether.**

**I would have very little quarrel with a quantum theory of atomic energy levels, were it extended to include somehow the phenomenon of external environmental energy and ether theory as it should have been logically construed, but I take issue with its present development, limitation, and abuse by the RQMs.**

**My position is that they already have covertly applied it to an ether theory, that this application is well concealed, and that this book is needed to begin the reconstruction of these concealed truths, for use by the public. While Planck said his quantum theory applied to a "system", he didn't say the "Relativist system of Einstein"—which came into vogue *after* Planck's theory—and was the only system it could be applied to. Since a rationally extended quantum mechanical theory would not exclude an ether, my "RQM" label distinguishes between the relativistic misinterpreters of quantum mechanics, and what could be called the "Integrated Reality Mechanics" ("IRMs").**

**The atomic hydrogen reaction first came to my attention in 1964, when I was studying industrial processes at Sam Houston State University, Huntsville, Texas, the year after taking an introductory course in college physics. While reviewing various welding processes in a textbook, my eyes fixed on an older process called "atomic hydrogen welding". By that time, the process was already considered "obsolete". To me, the process seemed valuable, not only because it produces such high temperatures—up to 5,000° C., more than enough to melt tungsten with a M.P. Of 3410° C., and the highest temperature producible by man—but also because it is "self-shielding", can be used to weld diverse metals (often without flux) and produces a concentrated flame, with little heat distortion when welding thin metal. In the process, 'normal' diatomic $H_2$ is shot through an electric arc which dissociates it into "atomic" hydrogen, $H_1$. This atomic hydrogen recombines at the (welded) metal surface,**

producing the very high heat. Though the process interested me then, and always has, I have never seen an atomic hydrogen welding unit for sale for the many years hence. Industry's obvious excuse for laying the valuable process aside was that it has been replaced by 'better' processes, such as Heliarc (TIG, "Tungsten Inert Gas") and MIG welding, though they also rarely mention "plasma arc welding", which has also almost disappeared from the market. Since plasma arc welding is merely an extension of the atomic hydrogen and TIG processes, using a specially redesigned torch, the mysterious reasons are undoubtedly the same. Neither of these two processes uses a lot of "consumables" (welding fluxes, rods, gases, etc.).

It was then that I got the idea of my atomic hydrogen furnace. The process simmered in the inner recesses of my mind, taking shape for a few years until 1976, when I rekindled my interest in the process for possible use in welding stainless steel and reducing and fusing platinum metal compounds, because hydrogen reduces such compounds (which must also be shielded from oxygen) to their metallic states. The atomic hydrogen process does not rely upon the combustion of hydrogen with oxygen in the air, but upon the purportedly "atomic" energy released when the atomic hydrogen recombines to form the 'normal' diatomic hydrogen. I still had some unanswered questions, since the various welding data at my disposal failed to mention sufficient specific details.

If Nikola Tesla was right, then I am right, that atomic energy comes through the ether from the Primary Solar Rays, not from the radioactive elements themselves.

Because I knew of no source from which to purchase an atomic hydrogen torch in 1985, I decided to build one, but my information was inadequate for proper construction. In the torch I made, the hydrogen entered the arc concentrically around both electrodes, instead of passing through the arc at a right angle. I also used the wrong kind of arc transformer, so it didn't work as well as it should have.

Despite the disappointment, I knew the torch would work better if I corrected the problems, so I kept the torch in my barn until better information and sufficient time was available. I excited the curiosity of the industrial spies when I rented a large tank of hydrogen at a local welders' supply, and this probably contributed to the sudden

**acceleration of the CIA-maintained judicial harassment that I endured between 1974 and 1992 (18 years). They apparently assumed I was exploring the process for its energy potential, rather than for just welding, and they were correct. My torch is shown below:**

**In the '70s, I had acquired an old English inorganic chemistry textbook[58], purchased from Los Alamos Scientific Laboratory salvage for $.25, which contained a halftone photo of a 1930s vintage atomic hydrogen welding torch, along with some basic data. The torch shown was different from the one I constructed, in that I used a high voltage arc transformer and had no mechanism to strike the arc electrodes together to start the arc, because the high voltage made it unnecessary. It also showed the gas passing through the arc at a right angle. In this old textbook, it was stated at page 170 (emphasis mine), as follows:**

***"Langmuir (1912) discovered that hydrogen at low pressure in contact with a tungsten wire heated by an electric current is dissociated to some extent into atoms:"..."This absorbs a large amount of energy, about 100 kcal. per gram-molecule." "...The atomic hydrogen formed is chemically very active. Atomic hydrogen is formed when an electric arc between tungsten electrodes is allowed to burn in hydrogen at atmospheric pressures (Fig. 106)."* (NOTE: This statement shows a basic misunderstanding of the process by Langmuir and others before my ideas emerged.) The**

58[58] J. R. Partington, *Text-Book of Inorganic Chemistry* (1953) Macmillan and Co., London.

**text continued:**

***"Atomic-hydrogen blown out of the arc by a jet of molecular hydrogen across the arc, forms an intensely hot flame, capable of melting tungsten (m. Pt. 3400°). This flame obtains its heat from recombination of hydrogen atoms to $H_2$."***

***"Hydrogen being set free in a chemical reaction is often more reactive than hydrogen gas."***

***"...the activity of such nascent (newborn) hydrogen, in the act of liberation from its compounds, is due to the hydrogen being in the atomic state."***

**The following is a copy of "fig. 106":**

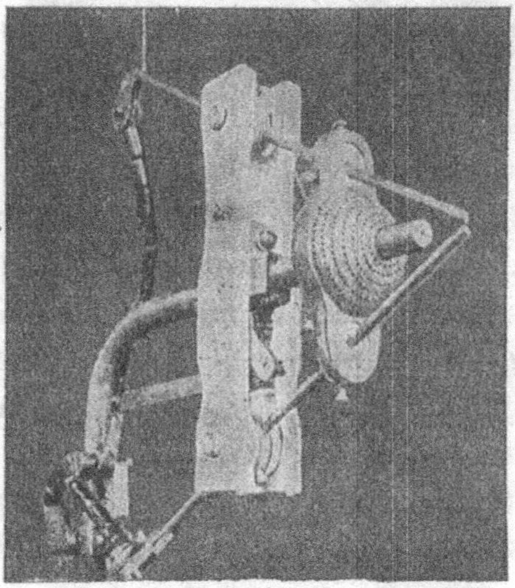

FIG. 106.—The atomic hydrogen blowpipe.

**In checking this data in a more recent Van Nostrand's Encyclopedia of Science[59], at page 1311 it was stated in pertinent part:**

***"Hydrogen molecules dissociate to atoms endothermically at high temperatures (heat of dissociation about 103 cal/gram mole) in an electric arc, or by irradiation." "...the hydrogen atoms recombine at the metal surface to provide heat required for welding."***

---

59[59]. *Van Nostrand's Scientific Encyclopedia*, Fifth Edition (edited by Douglas M. Considine), Van Nostrand Reinhold Company (1976).

**I was surprised to find this in the Van Nostrand encyclopedia, but I also found that in 1910, the D. Van Nostrand Company published a book by Haller and Cunningham entitled *The High Frequency Coil* (The construction of a Tesla Coil for the advanced amateur. 119 pages), so there is a record of the Van Nostrand company's association with Tesla technology.**

## SMOKESCREENS FROM ACADEMIA

**It seemed odd to me that it was later suggested that the 103 calories of dissociation energy absorbed from a very brief exposure to the arc is the same heat as that "...required for welding" as described, and I believed it to be more reasonable that the excess heat had to come from "elsewhere". I believe the true dissociation energy must be somewhere between the 103 calories and the 109 kcals./gram mole.**

**The older text showed the clearer construction of a device, and the newer text showed that only 103 cal./gram mole were required for dissociation, while the older text (Langmuir) showed that 100,000 cal./gram mole were liberated on recombination. Only by jumping back and forth between the two sources was I able to put a complete documentation together, and discover the conflicts. It meant that if the 103 cal./gram mole dissociation energy was correct, that with the 109 k.cal/gram mole output, the process was theoretically capable of producing 1,058 times the input energy. This is probably too high but it is definitely higher than all previous calculations.**

**It was apparent from the newer text that the writers intended for us to believe that the final 100 k.cal./gram mole heat—later upped to 109 k.cal./gram mole—was absorbed from the arc, but the 103 cal./gram mole dissociation heat figure showed a net 108,897 cal./gram mole unexplained. If there are about 22.4 cubic liters or 22,400 cubic cm. If that amount of hydrogen at STP can be dissociated by 103 calories of heat seemed astounding, but since hydrogen is the lightest element, it might be possible, but considering the idea that this same volume of hydrogen, weighing one gram, could somehow carry the heat of 109,000 calories—equivalent to the caloric heat of about 75 loaves of bread weighing at least 112 lbs.—it seemed ludicrous. This bread would weigh 50,400 times the gram of**

**hydrogen, and the actual hydrogen content would be so much greater than that gram of hydrogen, so that the "official story" seemed impossible. So the "extra heat" appears to come from an officially unidentified source.**

**The idea that sufficient energy to weld could be extracted by the arc during the time required for the gas to pass from the orifice and through the arc is remarkable. The time required for 22,400 cubic cm. to pass through the arc allows for .2 calories per cubic cm. Five cubic cm. of gas would be required to provide each calorie of heat, amounting to 1/9th of one slice of bread. A minimum number of cubic cm. per second of hydrogen involved in making a weld is obviously required.**

**109,000 cal./gram mole equals 432.5 BTU/gram mole—roughly the heat energy contained in 75 loaves of bread (1 lb. 8 oz. loaves at 90 calories per each of 16 slices)—so the heat energy which they have asked us to believe is 'stored' in an amount of atomic hydrogen which weighs 1/28th of an ounce, during its passage through the arc seems incredible. Yet the transformer is required to produce only half what it does in conventional welding processes. It seems more likely that excess heat could be stored in molecules than in 'almost naked' atomic hydrogen atoms. What ever happened to Bohr's little atom! It got bigger, and bigger, and .......**

**Between the older text (1921-1950, from the first and sixth editions) and the newer (1976) Norton science encyclopedia, it was obvious that science was much more simpler in the pre-National Security Act days, and that in the newer text, there was a reflection of the fishy attempts brought on by the national security laws and the firmer establishment of the RQMs, to conceal certain scientific knowledge which had been discovered all the way back in 1912 by Langmuir, and I even find it hard to believe that Langmuir did not know or suspect the truth. Aside from the discrepancies between the older measurements, regarding the comparable outputs between the combustion of gasoline and the combustion of molecular hydrogen, and the output from the atomic hydrogen process, to the newer text, there was a failure, so far as I could determine, to state under appropriate references (or any others I could find) some important facts concerning atomic hydrogen, and even apparent misrepresentations of the actual comparative outputs of gasoline and**

**hydrogen combustion, through "doublethink"—the Orwellian "badges of fraud"—of which I spoke in Space Aliens. Though this 103 cal./gram mole "slip" appears to have been made, Norton apparently was also forced by circumstances to rely on some 'scientists' who represented the fossil fuel cartels and other conflicting interests, while the older text, though inaccurate, should have been relatively free of such influences at that time, but maybe not.**

**For example, in the newer text, page 1311, energy from the combustion of hydrogen was stated at 29,000 calories per gram (52,200 BTU/lb.). In the older text, this value was given at 62,100 BTU/lb., with a comparative value for gasoline ("petroleum") at 19,800 BTU/lb., stretched to 20,825 BTU/lb. in the newer encyclopedia. These figures roughly comported with those in the encyclopedia for n-Heptane ("gasoline") at 19,314 BTU/lb., and hydrogen at 51,571.4 BTU/lb., more or less repeated at page 1137, with n-Heptane at 10,737.2 cal./gram, and hydrogen at 28,669.6 cal./gram, yet nowhere in the newer or older text was it affirmatively stated that the hydrogen was not "consumed" in the process.**

**The corrected figure (52,200 BTU/lb.) shows, in a comparison between the combustion of "normal hydrogen" (to form H2O) and gasoline (to form CO2), that hydrogen has a yield of roughly 2.7 times that of gasoline, by weight. Then we come to a comparison between gasoline and atomic hydrogen. Though the two can be compared BTU-for-BTU and pound-for-pound, showing atomic hydrogen's phenomenal output, the real measure of the atomic hydrogen process is gleaned only from a comparison between the "input energy)—103 cal./gram molecule—and the net output. The "input energy" would be "unity" for the process, because the process does not 'consume' the hydrogen, but only returns it to its associated diatomic state as $H_2$. The dissociation energy, subtracted from the gross output, would be the net output:**

**109,000 cal./gram mole (gross heat output)**
**Minus <u>103 cal./</u>gram mole (dissociation energy - "unity")**
**Leaves 108,897 cal./gram mole (net output - "over-unity" because the hydrogen didn't have that much calorific energy in the first place, and was not 'consumed' in the second place)**

This output is theoretically 1,058 times the input energy. In my opinion, hydrogen is not really a "fuel" in this process but rather a "medium" used in the extraction of and conversion of energy from the Primary Solar Rays via the ether, by transforming cosmic radiant energy into infrared (heat) radiation, which according to Tesla is actually (mechanical) sound waves in the ether. The energy required to pump the diatomic hydrogen to a holding tank, before being recycled and shot back through the arc and into the reaction chamber once again, is not considered in this computation because it is relatively negligible and unrelated to the process itself, since the dissociation energy is less than a thousandth of the gross output, and there might be a way to make the process work without this particular pump anyway.

The manipulators of information appear to have done everything they can to make a direct comparison between atomic hydrogen and gasoline as difficult as possible, but I have waded through a multi-step mathematical, physical, chemical and unitary process to get closer to the truth. According to 'official' sources, atomic hydrogen produces over 109 k.ca../gram mole, which is 109,000,000 cal./kilo. Subtract the 103 cal./gram mole endothermic dissociation energy, and a net of 108,897 k.cal./gram mole is left.

If true, the 1,058 times the input energy figure for atomic hydrogen however brings into question as to whether or not the official figures for gasoline are correct. Perhaps they are wrong, which would explain how Shell Oil Company in a battery of efficiency tests in the '50s were able to get hundreds of miles per gallon of gasoline in an automobile with a specially designed carburetor.

No wonder the atomic hydrogen welding process "went out of use". With this process, home consumers would pay nothing for fuel, because they could produce a small amount of hydrogen in the basement, any time needed, and could use it over and over, *ad infinitum.* They could heat their houses, drive their cars, and use it for "home industry" purposes. In a motor cruiser on the ocean, one would never have to worry about where the "next gallon" of fuel was coming from, or what it would "cost", because like sunshine, the heat from atomic hydrogen is "free", except, unlike sunshine, you could produce as much as you want, any time, day or night, summer or winter, rain or shine; the farmer would no longer be dependent upon the oil cartels for his energy, and could run his machinery and heat his barns "free". He could tell the fuel suppliers and the banks to screw themselves. People in the far Alaskan north could heat large spaces without huge expense. "Northern industry" would thrive again. The population control Nazis will hate me.

It is obvious that the reason that corporations like ARCO went so heavily into producing solar energy equipment in the early '70s, was so that they could control the patents and raise the costs of solar panels and ultimately control the cost of (free) solar energy, to make it "competitive" with gasoline and other fossil fuels. And they were eventually busted for this anti-competitive activity.

There are many ways used in contemporary texts to confuse facts, but the most effective way is by complete omission. Such facts could not have been concealed by "accident" or mere "oversight", because they are too phenomenal. Only by plowing through a tortuous mathematical process and confusing units and other factors was the true motive for such concealment revealed for certain, but tell me why I'm not surprised?

Conservatively speaking, the atomic hydrogen process bears a 10.5-to-1 ratio to n-Heptane. Would you care for 315 miles per gallon? How about 550,000 miles? It all depends on how many times you recycle it. You don't need to worry about wasting energy on electrolysis because you recycle the same hydrogen continuously.

The figures for n-Heptane are grossly confusing, as one might expect, since it is shown only by comparison by weight that hydrogen *combustion* has a 2.7-to-1 ratio to n-Heptane. This deception was effected by playing around with differing heat and weight units, pressures, and volumes, and the difference between the centigrade, Fahrenheit and English unit systems. The "109,000 cal./ gram mole" figure for atomic hydrogen was omitted from the newer texts, though one might have been able to extrapolate it from some complex data stated in other areas where even the inquisitive person is not expected to look. There was a total absence in the new text or other "new" data available to me, of a direct comparison of energy outputs between combustion of petroleum and hydrogen, much less the atomic hydrogen process. This is of course consistent with the continued government secrecy since 1943 regarding the huge mass of information on Nikola Tesla. Hide what you can. Lie, ignore or confuse about the rest. My invention has thrown all the data up in the air. Physics will have to make some fundamental changes.

I have never seen a direct comparison by any of the thousands of Ph.D.s in chemistry or physics, who should have become aware of these facts during their cursory studies. To make these figures even more ludicrous, with BTU ("British Thermal Units"), one would expect the weight to be shown in kilos under the metric system, instead of the English "pounds". These machinations appear to rely upon the well-founded assumption that all scientists who know, are controlled and don't say, and those who aren't controlled, will be considered "quacks" if they try to say. Again, it the usual case of "Those who know, can't say, and those who say, will be presumed not to know."

Another stupid trick used is to compare "system weights", under the false assumption that the user of hydrogen would have to use only the systems they present, raising the "comparable" weight of hydrogen to gasoline, using cumbersome cryogenic bottles or hydride storage tanks, etc., when the simple figures clearly show that one-third the amount by weight of hydrogen would be needed for a comparable combustion process. Whenever hydride systems are mentioned, there is an obvious omission of the liquid hydride system developed by Dr. Gerald Schafflander of California, the promotion for which Schafflander and associates were abusively prosecuted by the SEC, as previously stated in *Space Aliens* and *Pentagon Aliens*. Yet, the atomic hydrogen process would be more than adequate—even fantastic—with only a pound of hydrogen, and one could carry an extra 25 lbs of Dr. Schafflander's liquid hydride in the trunk. Below is a direct comparison between the BTU/lb for combustion of gasoline and molecular hydrogen, and the atomic hydrogen process, respectively:

| | |
|---|---|
| Gasoline combustion (n-Heptane) | 19,314 BTU/lb |
| Hydrogen combustion ($H_2$ + O) | 52,200 BTU/lb |
| Atomic hydrogen ($H_2 \leftrightarrows 2H$) | 196,200 BTU/lb |

Note that this comparison is based on the 'accepted' value for the

**atomic hydrogen process and does not involve a "consumption" of the hydrogen, yet even if it did, the ration between atomic hydrogen and gasoline is still 10.5-to-one. Take into consideration also that all the best methods for obtaining over a hundred miles per gallon (even hundreds of miles per gallon) of gasoline in automobile engines, utilize the vaporization of gasoline and mixture of it with air prior to combustion, something which is easier to do with hydrogen because it is in a gaseous state at normal temperatures and pressures. Also consider that while n-Heptane is a 'purer' gasoline, it hasn't been available for years, and most "gasoline" is a dilute mixture with water and various additives, with only about 50% n-Heptane, so I'm being as kind as I can be to "gasoline".**

**If a pound of gasoline could propel a car 30 miles, the consumption would be 666.6 BTU/mile. On the same scale, a pound of atomic hydrogen would yield 315 miles. Then you could repeat it, over and over and over. A pound of hydrogen could conceivably supply all your energy needs for your whole life, and you could produce another 100 pounds of it electrolytically, right in your basement, from tap water. Is this free energy, or what?**

**Though the facts of this "secret" hydrogen process are "hidden in plain sight", one must beware of the disinformation, as well as the common mistakes created by "scientists" who have been duped by the RQMs. For example, in the 1976 Norton Encyclopedia, 5th Edition, page 1311, most of the pertinent facts about hydrogen are shown. The "109,000 cal./gram molecule" for recombined hydrogen, reported in an older text, was not shown directly, though the following facts which were shown, are of particular note:**

| | |
|---|---|
| **Heat of vaporization at 20.4°K** | **107 cal/gram** |
| **Energy released upon combustion** | **29,000 cal/gram** |
| **Heat of combustion at 25°C (gross)** | **63,317 cal/gram** |
| **Heat of combustion at 25°C (net)** | **57,797/6 cal/gram** |

## ANOTHER ASTOUNDING PROOF AND MORE HYDROGEN METHODS

**The Norton encyclopedia was in its 5th edition, and the "103**

**cal/gram mole" dissociation energy did not appear to be a typographical error or misprint, and should have been corrected by that time if discovered. How many books would I have to search to find such a 'misprint', especially since I didn't expect to find the anomaly because of the obvious concealment? Yet, in Physical Chemistry[60], at page 418, appeared the following:**

***"The Spectroscopic evaluation of the dissociation energy***

***The energy of dissociation, $D_e$ of a diatomic molecule is the difference between the potential energy of the atoms when infinitely separated, and their energy when the atoms are at rest at the equilibrium separation."***

**Notice that the "at rest" state of the molecules is couched in language that makes you think that the "infinitely separated" atomic hydrogen atoms represents the "excited state" for the element. In layman's terms, the RQMs simply made the statement that the dissociation and recombination energies had to be the same, because their theories depended upon them being the same. Then they measured the dissociation energy backwards, by measuring the recombination energy, and stating the measurement as if it were the dissociation energy. The statements are supplanted by a lot of RQM gobbledegook, with a "spectrographic evaluation" thrown in like some snake oil, because the spectroscope purportedly "can't lie". There is also in the same text, at page 417, the following obtuse statement:**

***"...In kilocalories per mole, $D_e$ is 109."***

**If hydrogen atoms exothermically release energy when they combine to form molecules, the potential energy of the heat has been lost by the atoms, yet they attribute the "potential energy" to the hydrogen molecules—backwards—to evade their duty to draw the logical conclusion. If this energy cannot be obtained except by the prior injection of 109,000 cal/gram mole, then it is not "potential energy". This is shown by a graph ("fig. 4") purporting to show the**

60[60] E. A. Moelwyn-Hughes, Cambridge, *Physical Chemistry* (1965) Pergamon Press, London, page 418 (Supra.).

"potential energy of the hydrogen molecule in the ground electronic state." This clearly appears to be a misrepresentation by a Ph.D. in physics of the "Royal Society", that hydrogen in the "atomic state" is not in a ground state, but in an excited state. In this 1965 physical chemistry text from England, the heat energy released on recombination of the hydrogen atoms is given as 109,000 cal/gram mole, then misrepresented as equal to the "potential energy" of the "ground state hydrogen molecules" which have just formed. But the "ground state" of hydrogen should be the atomic state, with the electron at its lowest level of energy, since there appears to be no way the naked atoms could "store" the 109,000 cal/gram mole. Where, exactly would this energy "reside"? I could believe that the atom could capture the energy from the Primary Solar Rays on dissociation, but since the heat energy is released on recombination, the molecules lack the heat energy already released. The potential energy of the separate atoms is greater than that of the molecules, which I think is because the heat energy is released from the entrained ether and converted Primary Solar Rays—not from the hydrogen atoms—when they combine to form molecules.

Since original publication of this book in 1997-98, there have been at least four specific cases where, based on my original invention, at least four tests have shown over-unity results. These are:

1. D. W. Energy Research, LLC, Colorado Springs, Colorado, U.S. Patent #6,113,065, "Liquid Gasification Process", claiming an "energy amplification factor of about 1.9" (a patent based on a verbatim copy of my design).
2. Alexander Frolov, Faraday Labs, St. Petersburg, Russia, claiming an efficiency of over 200%, using an A.C.-heated tungsten filament. (http:/faraday.ru).
3. Jean-Louis Naudin, Paris, over 2,100% effocoemcu isomg a 50 Hz. D.C. pulse (created from Eurpoean 50 Hz. House current).
4. Italians, Andrea Rossi and Sergio Focardi---the "Rossi-Focardi LENR Device"---based on the Lyne Atomic Hydrogen Furnace, but concealed by a deceptive description and "secrecy"). This was announced Jan. 14, 2011, from Bologna, Italy. Rossi claimed an output of 12,400 Watts with an input of

**300 Watts (COP of 31).**

**Though Naudin says that his experiment was based on Langmuir's work, Langmuir never invented the use of a closed loop system, and never knew that the process was over-unity, which means his calculations were WRONG. That is my work. I also am the first that I know of to design an industrial grade atomic hydrogen furnace. Naudin used the same unit constructed by Frolov (using a large old metallic water-cooled rectifier tube) with a slight alteration of the dissociation current (from a conventional electrically-heated tungsten filament). Both Frolov and Naudin mention a Danish man named Nicholas Moller who plagiarized my work and my design and misrepresented to Frolov that it was his work. Frolov corrected this fraud.**

**Andrea Rossi copied my invention with minor variations. He dissociates the $H_2$ to $H_1$ then reacts it in a stainless steel chamber packed with nickel powder around which a water jacket removes the heat and converts water to low pressure wet steam. Then he recirculates the hydrogen circuitously as in my invention to be repeated *ad infinitum*. There is a difference in design but in effect no difference in concept.**

**The dissociation method used by Rossi---Nichrome® resistance heating wires---after which the $H_1$ is injected into the reaction chamber packed with nickel power, provides the environment which effects the recombination of the $H_1$ to $H_2$ with release of heat. This is said by some to produce some nuclear radiation as a result of energy-consuming transmutation of some of the nickel to copper (ca.10%) and iron (ca.14%), but there is no residual radioactive matter. The reaction chamber has a surrounding water jacket through which water is circulated to remove the heat and generate low pressure wet steam. The nickel buffets and lowers the high temperature reaction heat, so the water is of insufficient temperature to produce the dry, superheated steam required to power a turbo-electric steam generator. I solved the problem of how to handle the extremely high temperatures years ago but am now holding that secret as proprietary technology.**

**The transmutation effected by Rossi's device is not desirable with this process and is counter-productive, since it diminishes energy output and also leads to confusion for those who need a**

**conventional explanation. It has also confused Rossi, whose explanation has been rejected because it is a failed attempt to satisfy conventional physics. Only my "ether and Primary Solar Rays" explanation will suffice. Scientists must realize that with my invention, I have opened a door to use of cosmic energy. Until they realize this, they will only continue to prop up a false (relativist) physics, and incidentally continue to supply a market for uranium and nuclear energy, coal, gas and other fossil fuels.**

**There is a definite deception by Rossi, saying that he "invented" it and that he has a "secret catalyst" (when in fact, in this case, the catalyst is the "catalytic metal", nickel). This is shown in my original 1996 drawing as "*Activation energy device* (spark plug, *catalytic metal*, etc.)". The use of a catalytic metal is unnecessary so I apologize for that. Rossi's statement that 'there is absolutely no tungsten' in his device (in obvious reference to my design) only confirms his prior knowledge of my invention as well as confusion that my design somehow relies on tungsten, *per se* (it was recommended only for its high-temperature applications, not for catalysis). Almost any metal will trigger the recombination of $H_1$ to $H_2$. Once the reaction is triggered by contact with metal it will continue so long as $H_1$ is supplied.**

**I have also solved the exact mechanism by which the molecular binding energy is canceled and resupplied. I have also conceived a unique method for dissociation. But I won't disclose those details here because idea-thieves will only steal and say they discovered them.**

**My 1996 design is as follows (the encircled part is where Rossi got his "catalytic metal"/"eCat" idea):**

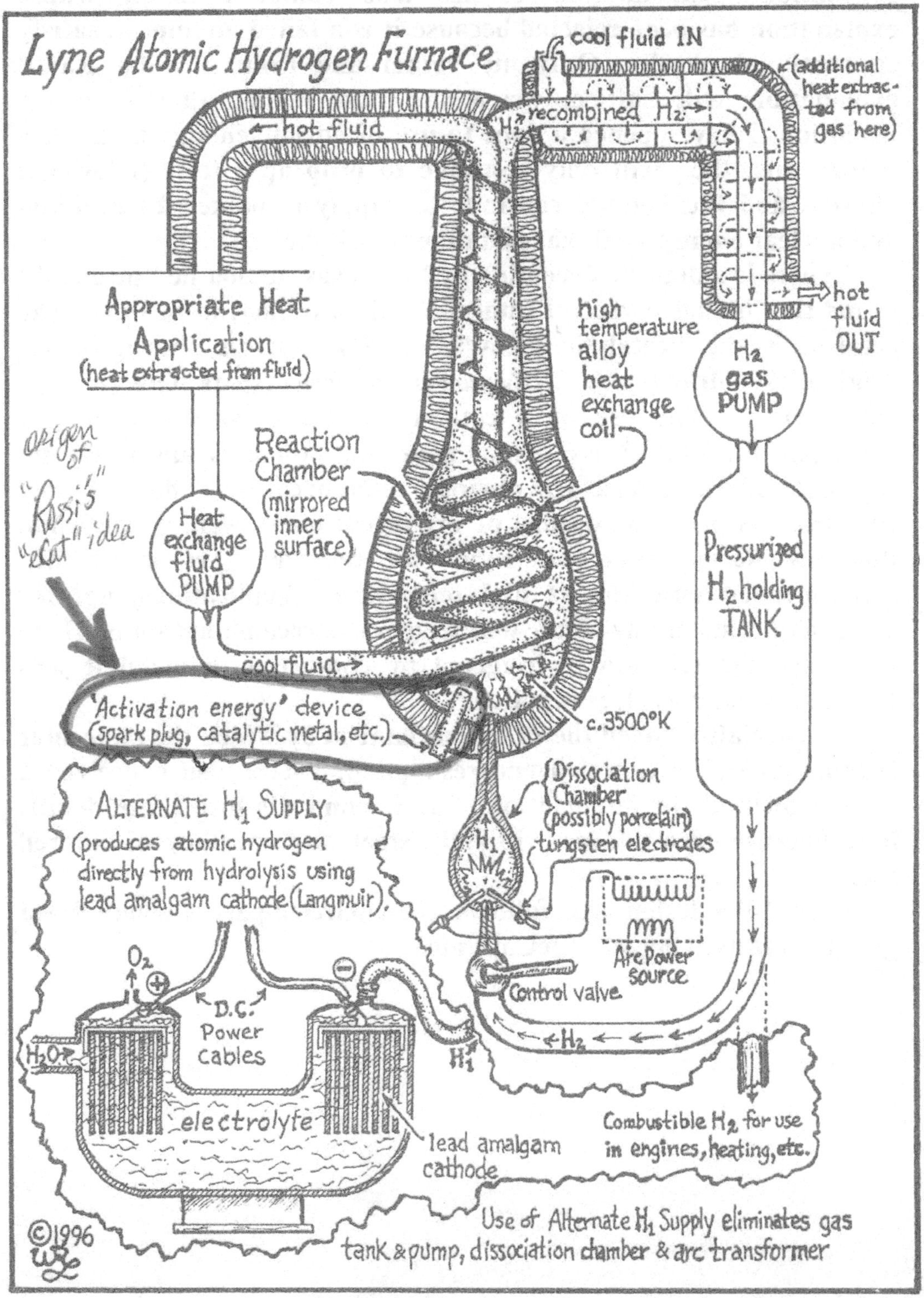
Lyne Atomic Hydrogen Furnace
cool fluid IN
(additional heat extracted from gas here)
hot fluid
recombined H2
H1
Appropriate Heat Application
(heat extracted from fluid)
high temperature alloy heat exchange coil
hot fluid OUT
H2 gas PUMP
origen of "Rossi's" "eCat" idea
Reaction Chamber (mirrored inner surface)
Heat exchange fluid PUMP
Pressurized H2 holding TANK
cool fluid
'Activation energy' device (spark plug, catalytic metal, etc.)
c.3500°K
Dissociation Chamber (possibly porcelain)
tungsten electrodes
ALTERNATE H1 SUPPLY
(produces atomic hydrogen directly from hydrolysis using lead amalgam cathode (Langmuir).
Arc Power source
Control valve
O2
D.C. Power cables
H2O
H1
H2
electrolyte
lead amalgam cathode
Combustible H2 for use in engines, heating, etc.
Use of Alternate H1 Supply eliminates gas tank & pump, dissociation chamber & arc transformer
©1996 WL

With both Langmuir's and Rossi's use of electrical resistance heating wires to dissociate the $H_2$, you have to ask what kind of radiation comes off the wires. I believe that it is not the heat---as Langmuir and Rossi believed---that dissociates the molecules, but the radiation, which according to Tesla is mechanical sound waves in the ether. These waves tear the $H_2$ molecules apart physically rather than by heat expansion. (Some have mistakenly described the $H_1$ as a "superheated hot gas" which is wrong.) The I.R. radiation is in the microwave spectrum. Other frequencies may work---some possibly even better---to get maximum effect, the idea being to use as little energy as possible to produce the greatest effect in the name of efficiency. The invention must also be durable and resilient to component fatigue, with minimal maintenance so as not to break down in the long Winters of Alaska or Sweden, or on a steam-turbine-driven ship in the middle of the ocean.

The following is a simplified drawing of my re-designed furnace, which does not include the secret details of each of the parts shown:

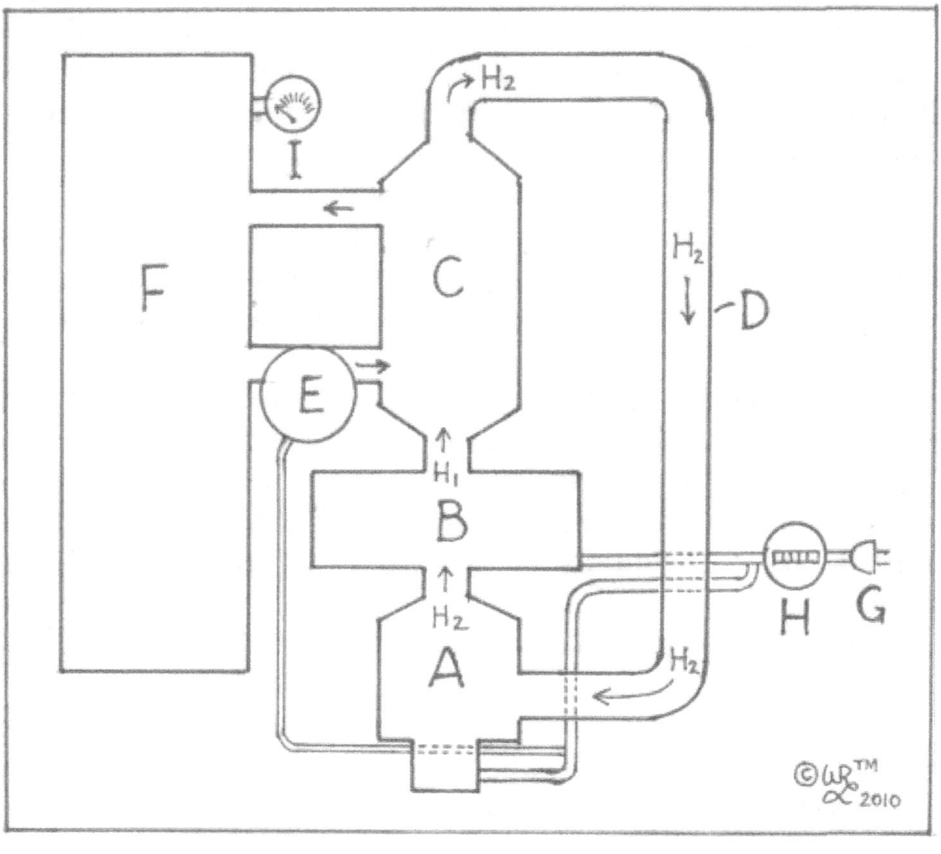

*A. - Accelerator*
*B. - Dissociator*
*C. - Reactor/Heat Exchanger*
*D. - H2 Return Conduit*
*E. - Heat Exchange Fluid Pump*
*F. - Hot Fluid Tank (test)*
*G. - Electric Input*
*H. - Watt-hour Meter*
*I. - Thermometer*

*H2 is accelerated by Accelerator A through Dissociator B where it is efficiently converted to H1. H1 is reacted in Combined Heat Exchanger/Reactor C returning H1 to H2. H2 is returned from C to A via H2 Return Conduit D. Heat Exchange Fluid Pump E pumps heat exchange fluid from Hot Fluid Tank F through Combined Reactor/Heat Exchanger C and back to Hot Fluid Tank F. Electric Input G Wattage is measured by Watt-hour Meter H and compared to calories measured by Thermometer I to determine efficiency ratio.*

**Inclusion of the details would enable the corporate thieves and other fakers to decipher my design, patent it, sell it to the government, and the people might never see it. The corporations would be unjustly enriched. A rough conservative estimate of this single unit's expected output, in a size no more than a cubic meter including the holding tank, is at least 50 kW net, free energy, and it will heat a large space or produce superheated dry steam to power a turbo-electric generator for electric power.**

**As of February, 2012, I have no backers, so building is slow on my meager means. Its secrets may die with me whenever I go, unless a fair backer emerges. At this point you might say that I am "on strike" like John Galt. But a pocket full of money will induce me to go off strike.**

**Due to changes in our patent laws---"first to file" not "first to invent"---I have decided not to open-source the details of my newest design, for then we might have the spectacle of some copy-cat suppressing me for authorizing the use of my own invention to people at large. The idiots in the Patent Office, in their infinite wisdom, probably thought this new rule would induce someone like me to apply for a patent. I do not have the financial resources to do so.**

**The RQM's misrepresentation of the real potential of this process is hereby unmasked as Orwellian doublethink. Since when is the potential energy of the atom of an element measured from its**

molecular state, or its atomic state called its "excited state"? This is exactly like saying that a truck sitting at the bottom of a hill has more "potential energy" than one sitting at the top, because of the energy required to drive it up to the top of the hill! What a strange way to define "potential energy". It is like saying that an empty gas can has the potential energy of the gas that it takes to fill it up!

The unique thing about the hydrogen atom, is that it is "born" at the top of the hill! Its astounding properties lie in its ability to perform as a "doorway" between the ether, Primary Solar Rays and the world of more familiar atomic matter.

The relativist statements have been unwittingly accepted as correct, because there is no ether or Primary Solar Rays possible under their theories, and the molecules have to have more energy than the atoms separately, because "$E = MC^2$". The RQMs have misrepresented and concealed this process in which separate atoms—with lower MASS—possess more "atomic" energy than the molecules!. These facts on hydrogen show that E equals MUCH MORE THAN $MC^2$.

Under relativism "solid" matter is thought erroneously to be the most "compact", "stored" form of "atomic energy", because the energy is supposed to be released from mass by the "smashing" and destruction, or fusing together of atoms, which Tesla said was false. It is clearly shown by the *Lyne Atomic Hydrogen Furnace* that Tesla was right and the relativists and Einstein were wrong, because the hydrogen atoms acquire more potential energy between the separate atomic state and the diatomic molecular state when the atoms recombine and the atoms have to get their energy from an external source. The only other way that the molecules can release any "potential" energy is through combustion with oxygen, which is a much lower energy output, and that too may come from the Primary Solar Rays. Tesla said that radioactivity was a reaction of atomic mass with the Primary Solar Rays, just as the hydrogen atom emits IR radiation. The separate atoms must convert kinetic energy from the Primary Solar Rays and eject it in the form of sound waves in the ether in the microwave spectrum when recombined. This radiation then reacts with matter to produce heat.

The perverse "potential energy" fallacy is also a way to confuse the diatomic hydrogen molecules with the nascent atomic hydrogen

**atoms produced by dissociation. The energy of hydrolysis is equated with the energy of dissociation, and the energy of combustion with the energy of recombination, yet the information presented here shows that by combining a type of hydrolysis of water, in which atomic hydrogen is produced directly, with the atomic hydrogen recombination process, instead of the combustion process, that at least 10.5 times as much energy as the hydrolytic input can be produced, in still a third astounding hydrogen free energy process. This is a gain in energy which Langmuir seemed to to have missed and failed to mention. This is true because the atomic hydrogen process without question releases at least 3.75 times the BTUs by weight (even by RQM standards), as ordinary hydrogen and oxygen produce in the combustion process. It is generally agreed that the "energy of combustion" equals at least the "energy of hydrolysis" (even though it is well known that the industrial production of hydrogen is generally 120% efficient in the so-called "water-gas" process, in which steam is shot through a white-hot graphite orifice).**

**Production of atomic hydrogen directly by hydrolysis, eliminating a separate dissociation operation, is yet another exciting possibility. If the 103 cal/gram dissociation energy is correct and the electrolytic energy is recovered by combustion, there is a theoretical possibility of getting a net 1,058 times the energy of input in the recombination process.**

## HOW DOES IT WORK?

**As a "welding process", this process was allegedly 'obsolesced' by MIG and TIG welding, neither of which can compare to its welding efficiency and uses. Since the welding process was its main use in the years past, the "new" processes allowed industry to toss the 'obsolete" process onto the junk heap. The "new" processes were promoted by the establishment, motivating their suppliers and the electric power companies at their own levels. More archaic tanks, transformers, gauges, torches, electrodes, gases, fluxes, and electric power, etc., etc., to be sold at a hefty profit. If people generate their own hydrogen for welding at home, how much equipment, supplies and power will they buy? Is saving money a crime?**

**How does the atomic hydrogen obtain its energy, if not from the**

**"ether" and Primary Solar Rays? No wonder establishment science doesn't want you to know there is an ether. If we are to believe the "law of conservation of energy" as interpreted by ether-excluding establishment "science", this process is impossible, yet using data available from the 'standard' texts, I have shown that the input energy of 103 cal/gram mole (or however much is required) is somehow 'magnified' by use of hydrogen as a 'medium' so that the 103 cal/gram mole 'seed' energy triggers the atomic hydrogen's apprehension of a net 108,897 cal/gram mole from the Primary Solar Rays via the "ether".**

**I reiterate, an equally phenomenal "atomic helium" process (reported in Space Aliens) is 4.6 times this caloric output per gram mole. What is the proportionate input? It can be conducted almost the same way, and produces the same kind of electrical energy and radiation-to-heat energy transformation, in which "mass" is only a "medium" used in the process. Similar principles are shared by the Lyne Atomic Hydrogen Furnace, as shown by my illustration at this chapter's end, and Josef Papp's noble gas engine[61]. If these processes used as much electrical energy as they produced, and electrical energy was the end product sought, what would be the reason for the process, right? At the comparable 30 mpg for n-Heptane, helium could produce 42.86 times as much as atomic hydrogen, or about 1285.8 miles per pound. That is, if you used the helium up, which you don't.**

**According to the Van Nostrand's Scientific Encyclopedia[62], simple electron bombardment of helium gas in a special chamber, excites the helium atom to a temporarily radioactive isotope of extremely short half-life. In this state, the helium atom unpairs its electrons, promotes one electron to the 2s state (in which state it remains for only a very short time), and afterward releases 460,000 cal/gram mole on collapse of the excited atoms back to their ground state. This gram of gas molecules occupies only slightly more than a gram mole of atomic hydrogen. It is my belief and theory that the elements in the excited state somehow convert the energy of the Primary Solar Rays according to Tesla's theory of radioactivity.**

---

61[61] Josef Papp, *Method and Means of Converting Atomic Energy into Utilizable Kinetic Energy*, Patent # 3,670,494 (June 20, 1972).
62[62] *supra.*

**Since each hydrogen atom is composed of a proton and an electron, each with equal-and-opposite charge, one would expect the binding force between diatomic molecules to be very weak, but when the atoms are separated they purportedly expand greatly in size, encompassing a greater volume of ether. On recombination, the atoms shrink in volume, and according to standard theory, infrared radiation is squeezed out of the atom. The dissociation requires only 103 cal/gram mole, yet on recombination, the atoms "exothermically" yield a net 108,897 cal/gram mole. How exactly does this occur?**

**All space is packed solid "bumper-to-bumper" with ether. This elusive and ultra fine, absolutely transparent matter is being omni-directionally interpenetrated by Primary Solar Rays composed of minute particles propelled at high speed by the enormous forces in stars. These particles must collide with and excite the ether held within the atomic space between the proton and the electron cloud, transferring their momentum to the ether and producing sound waves in the ether in the infrared (heat) spectrum, while the minute solar particles themselves must be absorbed by the atoms creating new mass or discarded. The increased momentum must be carried by the mono-atomic hydrogen atoms until they are opened up on release during recombination into diatomic molecules.**

**The facts suggested here are forbidden by relativity and quantum mechanics, in which a simple harmonic oscillator cannot have a stationary state of zero kinetic energy, because the ground state still has one-half quantum of energy and the appropriate motion, yet by relativity's 'other' rules, the principle of kinetic energy has no applicability to electromagnetic radiation. Since relativity also rejects the existence of an ether, there is nothing in so-called "empty space" to contain such radiant energy, and the atomic hydrogen process would have to create its energy from "nothing", therefore violating the law of conservation of energy, as misinterpreted by the relativists who see all energy functions as part of the 'winding down' products of an "original" BIG BANG and entrained entropy theory.**

**Incidentally, this same atomic hydrogen process, as first published here and now, is also the apparent source of the anomalous exothermic heat produced in aqueous cavitation, as well as in the so-called "cold fusion" process, which are two other free energy**

**processes which I believe are versions of the atomic hydrogen process.**

**The atomic hydrogen atoms have single, unpaired electrons in enlarged shells. These atoms are in Mendeleev's Group I-a, and all the atoms in that group have unpaired outer electrons, and are photo-reactive to and transmute when exposed to ultraviolet light, and all of them can be used to convert Primary Solar Rays into usable free energy. This photo-reactivity creates temporary, artificial-radioactivity-produced isotopes of short half-life, with the emission of radiant energy. But the radioactivity is contained within the atoms. The energy for the infrared frequency ether waves comes from the Primary Solar Rays according to Tesla, not from the atoms themselves. The atoms can be analogized by certain crystals, described by the Raman Effect, in which light passing through the crystals is "stepped down" to lower frequencies. The infrared spectrum light produced by the atomic hydrogen process is thermally reactive with normal atomic and molecular matter.**

**The transfer of force from the Primary Solar Rays to the dissociated $H_1$ atoms, is apparently the kind of thing spoken of by Tesla when he stated: "There is no energy in matter other than that received from the environment."[63]**

**In so-called "cold fusion", the electrolytic cell produces "nascent hydrogen", with the atomic hydrogen process, and the recombination of the mono-atomic hydrogen to $H_2$ releases over-unity heat, the product claimed. My process also proves Tesla's theory that atomic energy comes from the environment. It proves that "mass" does not even enter into the equation, and that Einstein's "E = MC2" is wrong. Einstein's theory says this process as I describe it could only be possible if the hydrogen were 'annihilated', converting its "mass" into heat, and otherwise, is impossible. But the hydrogen is still hydrogen, and Einstein was still wrong. It defies relativism and the BIG BANG.**

**Perhaps this or a similar process was also what Leland Anderson referred to in his eulogy of Tesla, when he mentioned his "plucking" thermal units "...right out of the thin air". After all, the hydrogen atoms when in the $H_2$ molecular form are smaller, therefore**

---

63[63] Nikola Tesla, *Lecture Before the Institute of Immigrant Welfare* (May 12, 1938).

"compressed". Compression of a gas releases heat radiation, usually thought to originate with the energy endothermically absorbed during decompression, which is backward from the gas dynamics. So in this case the question is "Where did the heat come from and exactly how much does it produce?"

In the case of hydrogen, the change is effected by the natural, physical process of recombination to $H_2$. According to standard theory, the molecular atoms can be 'decompressed' by application of "the heat of dissociation", because the strong binding force between the atoms is supposedly weakened by an amount of exothermic energy sufficient to separate the atoms, which is exactly the amount regained when the atoms are recombined.

Hydrogen atoms have the highest charge-to-mass ratio known, a ratio which almost completely neutralizes the protons. 'Decompression' of the electron clouds occurs during dissociation. As the protons dissociate with application of the 103 cal/gram mole, this allows the greatly expanded electron clouds of the atoms to encompass a larger volume of space, from which a net 108,897 cal/gram mole can be extracted from the only sources available, the Primary Solar Rays and the ether. Since there is no great cooling activity in the proximate area of dissociation, the 'heat' can only come from a conversion of environmental radiation, else there would be the formation of ice, etc. This process, though it can be thermally set up and triggered by electrons, is not merely a thermal process, because the net energy released is converted originally from kinetic energy, while the relativists have no plausible answer.

The raw data for this process came from ordinary science books. The full magnitude and output comparable to petroleum was only obviated by conversion of weights, measures, and temperature units, to produce a consistent picture and to obtain the total net outputs. Most physicists or chemists would refuse to believe the results, even if they had calculated them themselves, and would insist that they had erred. This is indeed an inverted BIG LIE which is told very small and often. The lie consists of a denial that there is an ether, the distortion of quantum mechanics, and the exaggeration of or minimizing of other incompatible measurements, to create a false overall picture. This is then backed up by a vacuum of information in texts and other sources available to the public, created via the

**National Security Act, at the behest of the Fuel/Power Cartels.**

## APPROPRIATE USE OF THE PROCESS

**With an understanding of the atomic hydrogen process, and the great magnitude of its output, what are its most immediately applicable uses?**

**The most direct, appropriate use of this technology appears to be to provide heat for homes, as well as for home-industrial or industrial heating purposes, or for agricultural heat production. A secondary use would be to produce heat for steam to drive turbine-powered electric generators or for motive power for cars, trucks, boats, trains and planes. This system would be excellent if coupled with refrigerant-driven, closed-loop turbo-electric generators, to provide electric power for electro-propulsive flying saucers, the ultimate in private transportation.**

**Since the net output is heat, the first order of business is a special heat exchanger, to transfer heat to water or other media, or to produce steam for turbine power. The heat in the reaction chamber of the atomic hydrogen furnace must be quickly removed by the exchanger to prevent a buildup of heat sufficient to melt the reactor surfaces which receive and catalyze the recombining gas atoms. The simplified design shown is merely a proposal which I presented to prospective backers without disclosing the true details.**

**Years ago, German metallurgists used atomic hydrogen in a copper crucible, creating a "hydrogen environment" to reduce platinum group metals from their compounds. To resist the high temperatures required for the reductions, the copper crucible was cooled from below by water, which removed the heat fast enough to prevent melting of the copper, as platinum metals were melted in it with the torch from above. The occlusion of oxygen prevented oxidation of the crucible and the platinum metals, and insured reduction of the platinum compounds by the hydrogen. This idea, together with over 45 years familiarity and used of the pyro-technology and over 34 years of welding has influenced the design of my atomic hydrogen furnace. The following is a simplified graphic showing the proposed circuitous cycle of the hydrogen atoms:**

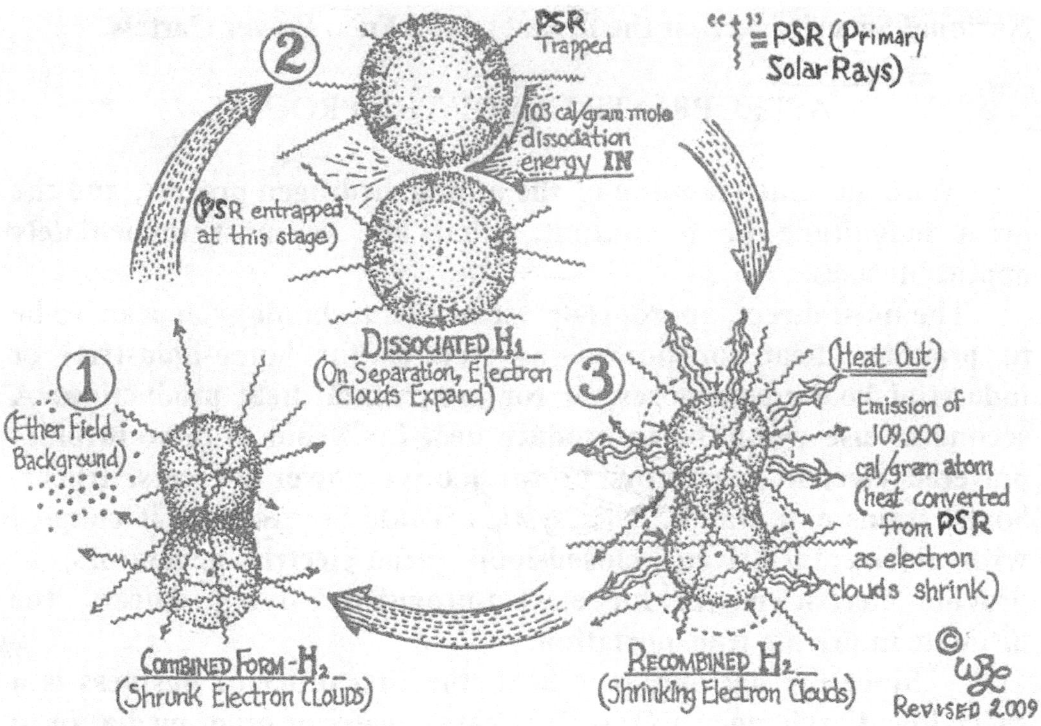


**At "1" above is the diatomic hydrogen molecule $H_2$, with more tightly bound electrons. At "2" are the dissociated hydrogen atoms, $H_1$, separated due to the introduction of dissociating energy which can come in several forms. At "3" are the recombined atoms, with the emission of I.R. frequency sound waves in the ether as the electron clouds shrink. The entry into the process by the Primary Solar Rays is indicated by "PSR".**

## CONCLUSIONS

**196,200 BTUs is a lot of energy from a pound of hydrogen. I have to pinch myself. Then repeat it many times. And this is only the tip of the iceberg.**

**Soon after this kind of "atomic energy" hit a small segment of the unknowing American public—the welders—in the late '40s, there were newspaper stories saying that the trans-Atlantic voyages of the Queen Mary could be powered by a hunk of radioactive material the**

size of a golf ball. They exaggerated their stories that way when the nuclear industry wanted us to get hooked and pay for it, but today's "nuclear reactors" produce electricity at the same efficiency level as coal and gas-fired generators (38%), and we still have to deal with the inevitable problems of greater expense, entrenched, archaic, stagnant "nuclear" and fossil fuel technology, nuclear decay, waste, and probable devastation like Cernobyl.

The stumbling blocks to the use of hydrogen are actually illusions, created by the corporate-state coercive monopolists. The atomic hydrogen effect was discovered in 1912 by Langmuir, and was developed only as a "welding process" in the '30s. The DOE 'scientists' at Los Alamos said the use of hydrogen involved heavy, cumbersome cryogenic tanks and equipment, until a liquid hydride system was developed by Dr. Gerald Schafflander of California, and was pointed out by me in a letter to the New Mexican newspaper in Santa Fe, New Mexico in 1980, totally ridiculing their phony efforts in 1979. Then Schafflander and associates were prosecuted by the SEC—because the liquid hydride stored hydrogen in something like "rock salt"—which cracked out at about 70° C. (The SEC presented this as incredulous, I don't.) DOE scientists mention only the heavier, more cumbersome, solid metallic hydride systems. Schafflander's lighter liquid hydride system, which used solar-voltaics to generate the hydrogen, would be a wonderful way to store and deliver hydrogen for the atomic hydrogen process. Goodbye oil cartels, hello atomic hydrogen, it's pay-back time!

By the way, the reason my "cold fusion" process (using my weird lead-based cathode) which I tested back in 1980, worked so well was because Langmuir, back in 1912, said his lead amalgam cathode produced "nascent" (atomic) hydrogen directly. Cold fusionists, throw away your palladium cathodes and get lead amalgam! Watch the price of water rise to $4.00 per gallon. After all, why should "they" allow "us" to buy a whole gallon of pure H2O "chemicals" for less, especially when a gallon is enough for a lifetime?

If the energy of hydrolysis produces approximately the same heat energy as that recovered by combustion, and dissociated hydrogen atoms can be produced directly by hydrolysis, this process with the furnace has to be free energy.

**William Lyne was born in Big Spring, Texas in 1938 and was raised in West Texas oil boom towns, with two years as a child in the Bay Area of Northern California during WW II, and the southeast Big Thicket country of Texas. He has a B.S. in Fine Arts and Industrial Arts (1965) from Sam Houston State University in Huntsville, Texas and an M.F.A. in Studio Arts (1969) from the University of Texas at Austin. Some of his family roots in Texas go back to 1820, sixteen years prior to the Texas Revolution. He has been living in New Mexico since 1970, and presently resides just to the south of Santa Fe, in the little railroad town of Lamy.**

**In 1993, he published *Space Aliens From the Pentagon: Flying Saucers are Man-Made Electrical Machines.* In 1995, a revised and expanded second edition of this book was issued. In 1997, the first edition of *Occult Ether Physics* was published. In 1998 the Second Revised Edition was published. In 1999 *Pentagon Aliens* (a third edition of *Space Aliens From the Pentagon*) was published. In 2002, Lyne published *Occult Science Dictatorship: The Official State Science Religion and How to Get Excommunicated.* In 2010 the 3rd revised edition of *Occult Ether Physics* was published in Russia.**

**As a result of these books, a vast quantity of false government propaganda has been published via covert intelligence agents posing as members of the "UFOlogy" community.**

**Lyne has been featured on such T.V. programs as the History Channel's "UFO Hunters" show (originally aired on April 22, 2009, pertaining to Nazi UFOs). In late 2011 Lyne was interviewed for a segment of Jesse Ventura's "*Conspiracy Theory*" TV show, which had not been aired by the time of this writing. Lyne has been interviewed several times by various radio talk show hosts, especially by his friend Erskine, of "Erskine Overnight", of Phoenix. Look him up and tune in!**

**Here's an illustration of a ship which Lyne saw in late summer, 1953, in Kermit, Texas while my family and our neighbors were having an ice cream party:**

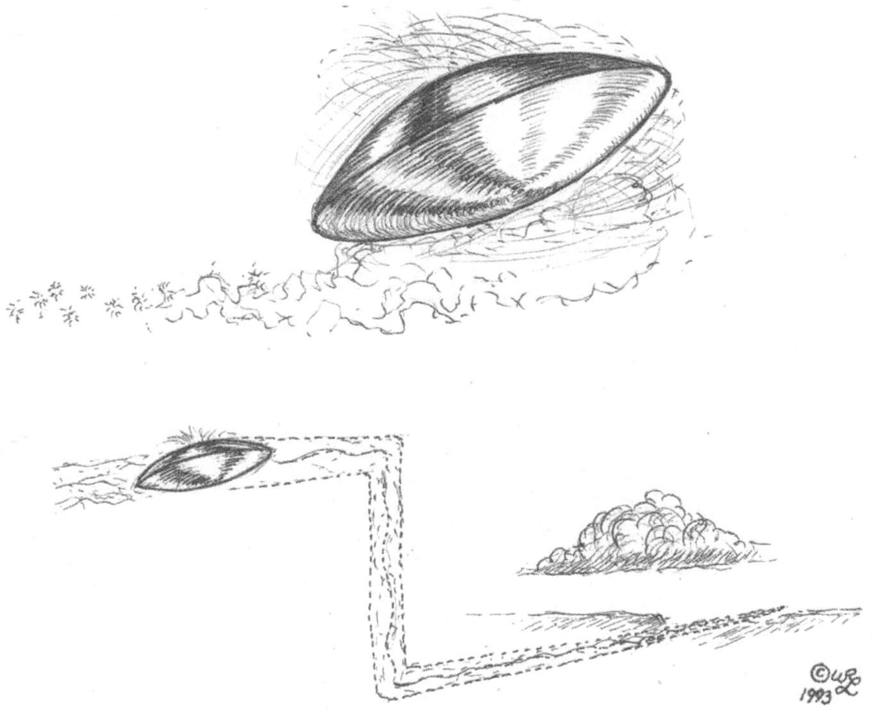

Made in United States
North Haven, CT
06 November 2026

21876160R00083